Tensor Product Model Transformation in Polytopic Model-Based Control

AUTOMATION AND CONTROL ENGINEERING
A Series of Reference Books and Textbooks

Series Editors

FRANK L. LEWIS, Ph.D.,
Fellow IEEE, Fellow IFAC
Professor
Automation and Robotics Research Institute
The University of Texas at Arlington

SHUZHI SAM GE, Ph.D.,
Fellow IEEE
Professor
Interactive Digital Media Institute
The National University of Singapore

Automation and Control Engineering Series

Tensor Product Model Transformation in Polytopic Model-Based Control

Péter Baranyi
Yeung Yam
Péter Várlaki

CRC Press
Taylor & Francis Group
Boca Raton London New York

CRC Press is an imprint of the
Taylor & Francis Group, an **informa** business

CRC Press
Taylor & Francis Group
6000 Broken Sound Parkway NW, Suite 300
Boca Raton, FL 33487-2742

First issued in paperback 2017

© 2014 by Taylor & Francis Group, LLC
CRC Press is an imprint of Taylor & Francis Group, an Informa business

No claim to original U.S. Government works
Version Date: 20130613

ISBN 13: 978-1-138-07778-2 (pbk)
ISBN 13: 978-1-4398-1816-9 (hbk)

Library of Congress Cataloging-in-Publication Data

Baranyi, Péter, 1970-
    Tensor product model transformation in polytopic model-based control / authors, Péter Baranyi, Yeung Yam, Péter Várlaki.
        pages cm. -- (Automation and control engineering)
    Includes bibliographical references and index.
    ISBN 978-1-4398-1816-9 (hardback)
    1. Intelligent control systems--Mathematical models. 2. Automatic control--Mathematics. 3. Tensor products. I. Yam, Y. (Yeung) II. Várlaki, Péter. III. Title.

TJ217.5.B37 2013
629.8'95630151563--dc23   2013018065

Visit the Taylor & Francis Web site at
http://www.taylorandfrancis.com

and the CRC Press Web site at
http://www.crcpress.com

Contents

Preface

This book introduces a methodology to numerically generate a class of invariant, canonical, and convex polytopic representations from any quasi-linear parameter-varying models. The class of polytopic representations is then readily executable for linear matrix inequality-based system control design. Our intention is to devise a standard design process that reduces analytical derivations as much as possible, echoing the recent paradigm shift toward the ready acceptance of numerical solutions as a valid form of output from control system problems. Our methodology is based on an extended version of singular value decomposition applicable to hyper-dimensional tensors. The resulting representations may exactly or approximately duplicate the original dynamics. Trade-offs between approximation accuracy and computational complexity can be performed through the singular values retained in the process. The book also proposes to manipulate the convexity of the resulting polytopic representations as a possible way to influence the subsequent linear matrix inequality-based design and control performance. This differs from the prevailing perspective in the literature, which focuses mainly on the proper formulation and conditioning of linear matrix inequalities to facilitate feasible designs. On the practical side, the book provides details about the application of the proposed methodology to design control systems with a number of real-life examples, including the aeroelastic wing section and the heavy vehicle rollover prevention problems. The book also introduces *TP-tool*, a MATLAB™ Toolbox which contains the relevant algorithms of the proposed methodology. The Toolbox is available online for free downloading.

This book is not intended to be a textbook, but rather a reference book for graduate students, researchers, engineers, and practitioners who are dealing with nonlinear systems control applications. The book will be a practical tool to systematically generate controller design for a large class of nonlinear systems, especially with the help of the MATLAB Toolbox *TPtool*. More importantly, we sincerely hope that readers will also find the book stimulating and useful as a platform upon which new concepts and MATLAB functions may be defined and explored as needed for analyzing their control problems and carrying out the design processes at hand.

The authors owe the completion of this book to a number of people. We give our appreciation to our young researchers Dr. Béla Takarics, Dr. Stephen C.T. Yang, Patricia Gróf, and Péter Galambos for their help in conducting experimental case studies and developing the MATLAB Toolbox; to Dr. Pauline Lee for her help in English editing; and to our graduate students, Mr. Kerney Wu and Miss Heran Song, for proof-reading parts of the initial drafts. The content presented in this book is based on our works over many years in the areas of fuzzy approximation, multidimensional singular value decomposition, and tensor product polytopic structure and control. We would like to thank our collaborators and graduate students, past and present, for their input and contributions to the various subjects. We also would like to thank the agencies who funded our research projects over the years. They include, especially, the Hungarian National Development Agency (ERC-HU-09-1-2009-0004, MTASZ-TAK) (OMFB-01677/2009) and the Hong Kong Research Grant Council (Project Ref. 2050520, 2150750). Part of the research was also supported by the Janos Bolyai postdoctoral scholarship.

Finally, we thank our families for their support during the writing of this book.

Péter Baranyi, Yeung Yam, and Péter Várlaki

MATLAB™ is a registered trademark of The MathWorks, Inc. For product information, please contact:

The MathWorks, Inc.
3 Apple Hill Drive
Natick, MA 01760-2098 USA
Tel: 508 647 7000
Fax: 508-647-7001
E-mail: info@mathworks.com
Web: www.mathworks.com

Acronyms and Abbreviations

CHOSVD	Compact Higher-Order SVD
CNO	Close-to-Normality
DoF	Degrees of Freedom
GEVP	Generalized Eigenvalue Minimization Problem
HOOI	Higher-Order Orthogonal Iteration
HOSVD	Higher-Order Singular Value Decomposition
ICA	Independent Component Analysis
INO	Inverse Normality
IRNO	INO and RNO
LFR	Linear Fractional Representation
LMI	Linear Matrix Inequality
LMIs	Linear Matrix Inequalities
LPV	Linear Parameter-Varying
LTI	Linear Time-Invariant
NN	Nonnegativeness
NO	Normality
PDC	Parallel Distributed Compensation
qLPV	quasi-LPV
qNN	quasi-NN
qSN	quasi-SN
RHOSVD	Reduced Higher-Order SVD
RNO	Relaxed Normality
SN	Sum Normalization
SMC	Sliding Mode Control
SMO	Sliding Mode Observer
SVD	Singular Value Decomposition
TORA	Translational Oscillations with an Eccentric Rotational Proof-Mass Actuator
TP	Tensor Product
TP model	Finite Element TP Type Polytopic Model

Chapter 1

Introduction

1.1 An overview

The book deals with two main areas of emerging importance in the field of systems and control, namely, the quasi-linear parameter-varying (qLPV) models and the linear matrix inequality (LMI) design method. The qLPV models efficiently treat nonlinear plants as linear time-invariant (LTI) systems with varying parameters that are functions of the state variables. Such formulation is applicable to a wide range of problems, for example, in the areas of aerospace control and gain scheduling. The LMI, on the other hand, is a very efficient computation-based method for multiobjective control system design. The integration of the two has been the subject of many ongoing works in recent years. This book, however, deviates from past approaches by bringing forth a number of novel concepts and perspectives.

Instead of using the given qLPV model directly, the book calls for the derivation and utilization of its finite element tensor product (TP) type polytopic model, or TP model for short, for design. Moreover, given that analytically deriving such representation will be a tedious or even impossible task, particularly for more complex systems, the book introduces a higher-order singular value decomposition (HOSVD)-based computational technique to numerically generate from any given qLPV model a TP model that is readily executable for the LMI control design. The TP model as generated may exactly or approximately duplicate the original dynamics, depending on the singular values retained in the process. In turn, the singular values serve as a measure to trade off the accuracy and complexity of the resulting model. For the subsequent LMI design process, the book adopts the parallel distributed compensation (PDC) framework, which uses the same TP form for the controller to readily incorporate multiple design objectives. The book also focuses on the generation of

1

advantageous TP models with proper convexity properties as a means to facilitate controller designs. This so-called convex hull manipulation for design is a new concept introduced in this book.

The book also introduces a MATLAB Toolbox called *TPtool* specifically developed for the present TP model generation, convexity manipulations, and LMI-based controller design. The book also includes a number of practical examples to demonstrate the application of the proposed computational-based modeling and design approach.

The following sections contain more detailed descriptions of the salient features of the book.

1.2   TP model

In the Paris Conference of the International Congress of Mathematicians in 1900, D. Hilbert gave a famous speech culminating in the publication of 23 conjectures on unsolved problems, which he believed would provide the biggest challenge in the 20th century. Specifically, he hypothesized in the 13th conjecture that there exist continuous multivariable functions which cannot be decomposed as a finite superposition of continuous functions of a lesser number of variables [GG00], [Gra00], [Hil00], [Kap77]. This hypothesis was disproved by Arnold in 1957 [Arn57]. In the same year, Kolmogorov [Kol57] formulated a general representation theorem, along with constructive proof, to express continuous multivariable functions in terms of one dimensional functions. This proof justified the existence of *universal approximators*. Kolmogorov's representation theorem was further improved by several authors, including Sprecher [Spr65] and Lorentz [Lor66]. With these results, it has been proved from the 1980s that the approximation tools of biologically inspired neural networks and genetic algorithms are all valid universal approximators, and the same is true for the inference-based fuzzy logic mappings [BL91], [Cyb89], [HSW89], [Cas95], [Kos92], [Wan92], [NK92], [EHR94]. As such, these approximation tools have been widely adopted in the identification of system models. They turned out to be particularly useful for approximating systems that may not be describable by analytical equations.

The above approximation tools thus add to the number of effective identification techniques that we have today. However, the identified models they produced are quite different in form and description to the usual models derived from analytically closed formulas via physical considerations of a given system. For instance, the neural network model would be a graph with a huge numerical array of connection weights, or the fuzzy logic mapping would be a set of linguistic rules. Furthermore, some identification techniques would represent the model in the form of algorithms, as in the genetic algorithm-based techniques. We can also have black-box methods which produce numerically executable computer simulation models, as in the MATLAB-Simulink Toolbox. We can also have models that are generated by mixing the various submodels given by the different techniques above. This requires a formulation to effectively handle this assortment of identification techniques and conceptually different types of identified model descriptions.

This book proposes the TP model as a possible solution. The model represents a multivariable tensor function as linear combinations of products of vector functions of single variables. Transformation into the TP model form can be executed regardless of whether the identified model is given in the form of analytical equations resulting from physical considerations, or as an outcome of soft computing-based identification techniques such as neural networks or fuzzy logic-based methods, or as a result of a black-box identification process, without analytical manipulation and within a reasonable amount of time. The only requirement is that the identified model must be discretizable over a grid. The TP model provides a uniform representation equally attainable from the outcomes of various identification techniques, and it is also readily applicable for modern control design processing.

1.3 HOSVD-based computation

Following the discussion above, analytical conversion of a general qLPV model into the TP model form is an extremely tedious process, if achievable at all. This book introduces a computational process based on HOSVD to numerically convert any given qLPV model into such form. Singular value decomposition (SVD) of matrices is one of the widely applicable tools in the field of linear algebra and linear algebra-based signal processing. The history of matrix decomposition goes back to the 1850s [Ste92]. During the last 150 years, several mathematicians, like Eugenio Beltrami (1835–1899), Camille Jordan (1838–1921), James Joseph Sylvester (1814–1897), Erhard Schmidt (1876–1959), and Hermann Weyl (1885–1955), as a few of the more important ones, made significant contributions in formulating the theory of the method. The pioneering works of Gene Golub further established the existence of efficient and stable algorithms for its computation [GK65]. More recently, SVD started to play an important role in various scientific fields [Dep88], [Vac91], [MDM95]. Its popularity also grew with the availability of more and more efficient numerical methods. Moreover, the tremendous advancements in computing technology in recent years make handling of larger-scale, multidimensional problems possible, leading to a greater demand for the higher-order generalization of SVD for tensors. High-dimensional SVD was introduced in fuzzy approximation [Yam97, YBY99] and independent component analysis (ICA) [DLDMV94], as well as the dimensionality reduction for higher-order factor analysis-type problems to reduce computational complexity [DL97], to cite just a few. A detailed study of the theories and properties of multidimensional SVD was published in 2000 [DLDMV00]. Application of HOSVD on a given N-dimensional tensor yields a set of two-dimensional orthonormal matrices, and expresses the importance of the various components in descending ordering of singular values.

The HOSVD-based computation introduced in this book extends the application of HOSVD to continuous multivariable function in the qLPV model, and numerically produces its TP model. The process, termed HOSVD-based TP model transformation here, produces a set of LTI vertex systems and a set of weighting functions of each dependent variable. Note that this structure may not be analytically derivable as there may not be a general analytic solution for the HOSVD, or that component(s) of the original qLPV model may be identified using a black-box identification process. In any case, the HOSVD computation has the advantage of replacing the need for analytical derivation, which in many cases is complex, nontrivial, or even impossible, with numerically tractable and straightforward operations that can be carried out in a routine fashion.

Similar to the tensor case, the importance of the vertex systems and corresponding weighting functions in the TP model are not equally important. Their importance are indicative by the associative singular values in the process. Keeping more singular values produces a TP model of enhanced complexity and also closer approximation to the original qLPV model, and the converse is also true. Actually, the error of approximation here is bounded by the number of discarded singular values in the process. The HOSVD process hence provides a tool for trading-off between complexity and accuracy of the resultant canonical form, which is another salient point in this book.

Moreover, in-between the original qLPV mode and the extracted TP model, the HOSVD computation also generates as a special form under which the associative weighting functions are orthonormal to each other. While this so-called HOSVD-based canonical form does not satisfy the convexity condition and is not used in the ensuing control system design, it does possess some interesting properties that are worth mentioning in greater detail in the book.

1.4   Convex optimization via LMIs/PDC framework

This book adopts the use of LMI-based methods under the PDC framework for controller design. This combination provides us with the following advantages:

- We can be assured of a solution via convex optimization given a set of feasible linear matrix inequalities (LMIs). The use of convex optimization in practical applications was made possible by the introduction of *interior point methods* since the 1980s. These methods were developed in a series of works after the polynomial-time projective algorithm by Karmarkar [Kar84a], [Kar84b]. They focused mostly on linear, and to a lesser extent, quadratic programming problems. Comprehensive extension of the interior point methods to solving convex problems, including the LMI, was provided by Nesterov and Nemirovskii in their book [NN94]. Convex optimization is deemed an effective tool in solving "everyday" engineering problems today.

- The LMIs are capable of incorporating multiple performance objectives in the optimization problem. Starting with the pioneering works of Gahinet, Bokor, Chilai, Boyd, and Apkarian (see [AGB94], [FAG95], [NG94], [Sch92], [DGKF89], [Gah94], [GL94], [AG95], [BBK89], [KKR93]), that casted *Lyapunov-based stability criteria* as LMIs, it was soon established that problems such as optimal LQ control, H_∞ control, minimal variance control problems, robust stability, pole placement, input/output constraints, and so forth, can also be efficiently converted (see [EB02], [SBS99], [GB99]). The LMI-based formulation also covers other areas such as estimation, identification, optimal design, structural design, and matrix-sizing problems. One can hence readily combine different system and controller design specifications and objectives together in the form of a numerically manageable set of LMIs [BEGFB94].

- The PDC framework facilitates an efficient basis for control system design. The framework was initially proposed by Tanaka and Wang in 1995 [WTG95] and later extended to handle multiobjective control design specifications in [TW01]. In particular, the structure of the PDC framework allows direct application of the TP model more readily than other types of LMI-based design formulations. Specifically, the framework calls for designing a controller adopting the same weighting functions as the TP model and determining the vertex control gains from the set of vertex systems via LMI-based optimization. Such an approach provides a simplified structure for the establishment of a feasible design with required performance, as weighting functions are excluded from the process. We will also have the advantage of using a number of theorems and results from previous works on PDC for this purpose.

1.5 Model convexity and convex hull manipulation

The TP model resulting from HOSVD transformation is actually not unique, that is, while the input-output relation remains invariant, the type of convexity of the TP model can be varied. This book defines a number of convexity types and presents various algorithms to incorporate different types of convexity into the TP model. During the controller design process, it is further recognized that the convexity type of the TP model will have strong ramifications on the performance and even feasibility of the resulting controller. It may be advantageous to tune the convexity for enhanced performance. This convexity tuning turns out to be equivalent to manipulating the convex hull formed by vertex systems of the TP model within the dependent parameter space. In other words, in addition to relying solely on the formulation of more effective LMIs for control system design, which is the widely adopted approach in existing LMI-related studies, this book calls for a systematic modification and reshaping of the polytopic convex hull as well. This is a key feature of the book that has the potential to set forth a new direction in control system design.

1.6   Significant paradigm changes

Computational-based approaches are becoming effective tools in the analysis, design, and simulation of engineering problems due to the tremendous advances in computing capability of the past decades. This brings about a paradigm shift on the acceptance of the numerical solution as a valid form of output to control system problems. A notable example is the LMI-based design as mentioned above, whereby a host of control problems are converted into convex optimization forms and solved with extreme efficiency. Nowadays, a control problem can actually be considered as good as solved if it can be put into LMI form.

One of the goals of this book is to advocate the development of computational-based solutions to control problems. We aim to establish a set of nonheuristic and standard computational procedures to automate the control system design process with minimal human interaction and tedious analytical derivation. Toward this end, the HOSVD-based TP model transformation and LMI-based control design methods adopted in the book are tailored to be numerically compatible with each other. The eventual output will be a numerical solution that "computationally" satisfies the requirements for guaranteed performance. A major advantage of such an approach is that in the case of a revision in the system model parameter, structure, or specifications, one needs only to rerun the algorithms to obtain a new solution. The standard methods that depend on analytical derivation, however, will take a long time to go through the process even if the revision is a minor one. Another advantage of the computational approach is the possible tackling of a wide class of qLPV control problems, whether the model is described in analytical equations, neural networks, fuzzy rule sets, or other soft computing black-box algorithms, or a combination of them.

The algorithms presented in this book are included in a MATLAB Toolbox called *TPtool*, which readers will be able to download. The book strongly encourages readers to try the algorithms on their own problems. Along this spirit, the book has also left much room for readers to explore. For example, the book defines and evaluates a number of TP model convexity types but the list is by no means exhaustive. Readers are encouraged to define and experiment new convexity types that suit their problems. Furthermore, instead of aiming at the fine details and mathematical conclusions pertaining to a particular control formulation and design methodology, the book actually hopes to illustrate how freely one can readily create new possibilities and MAT-LAB functions as needed for analyzing the problem at hand in a relatively short time. In this light, the PDC framework within LMI-based design adopted in the book can be deemed merely as an efficient illustration of the end-to-end computational-based design process. Readers are encouraged to explore different formulations. Hence, the book should be used as a platform free for others to explore and try out, and not as a fixed set of algorithms to be used "blindly" on any control design problems.

1.7 Outline of the book

The book is divided into three parts. Part I, containing Chapter 2 to Chapter 9, introduces the basic concepts of the TP model and its associated transformation and numerical manipulations. Chapter 2 presents the definitions of the TP model and basic convexity types. Chapter 3 gives the HOSVD-based computation for converting a qLPV system into the TP model form, including discretization of the qLPV model over a domain of interest, the application of HOSVD to the sampled data tensor, and the determination of the TP model structure and components. Chapter 4 defines the HOSVD canonical form of the qLPV model. Specifically, it shows the exact numerical equivalence of both in the limit of infinitely dense discretization. The chapter also includes the translational oscillations with an eccentric rotational proof mass actuator (TORA) system as an example. Chapter 5 presents the case to adopt a TP model that is not the exact equivalent of the original model as a trade-off study between the accuracy versus complexity of the ensuing model. A number of examples, including the TORA system, are included for illustration. Chapter 6 presents the transformations to incorporate different types of convexity into the TP model, a maneuver that will have strong implications on the control system design. To facilitate the computation of the TP model transformation, Chapter 7 introduces the MAT-LAB Toolbox *TPtool* developed specifically for the needed computation relating to TP model transformation and manipulation. The toolbox is available for downloading by interested readers at "http://tptool.sztaki.hu". Chapter 8 further presents the centralized TP model form which has several advantages over the standard one in numerical stability and precision. Chapter 9 presents the algorithms to conduct the HOSVD-based TP model transformation distributively, rather than centrally, in order to relax the computational load. This is important in the case of highly complex systems with a large number of dependent variables. An analysis of the reduction in computation load is also included.

Part II contains the next three chapters and is dedicated to TP model-based control system design. Chapter 10 outlines the different design steps involved and shows the initial step with the TORA system as an example. Chapter 11 gives a brief introduction to LMIs to show how they are utilized for control system design. It shows that the TP models can be readily fit into the PDC framework to facilitate LMI-based control system design. The chapter also shows how multiple control objectives can be incorporated into the LMI formulation. LMI-based observer design is also discussed. For illustration, the TORA example of Chapter 10 is followed up here with the derivation of a state feedback and an observer-based output feedback controller. Chapter 12 focuses on the importance of shaping the convexity types of the TP model, which amounts to manipulating the convex hull of the TP model, in order to attain the "best" possible control system performance. This underscores the importance of convex hull manipulation as a new direction in control system design not covered before in previous literature.

Part III serves to demonstrate the application of the TP model-based approach for computational control system design with a number of practical examples. Chapter 13 first gives a brief description of the LMI-based design functions in the *TPtool* toolbox. Chapter 14 presents the controller design of a two degrees of freedom (2-DoF) aeroelastic wing section with structural nonlinearity. Both state feedback and observer-based output feedback cases are generated. The chapter has also included discussion on convex hull manipulation and its effect on the controller performance. Chapter 15 considers the control problem for the three degrees of freedom (3-DoF) version of the aeroelastic wing section and shows the readily efficient extension of the TP model design approach to arrive at an acceptable solution. Chapter 16 presents the control design of a 3-DoF helicopter with four propellers subjected to dynamic uncertainties and design specifications. Finally, Chapter 17 deals with the case of a heavy vehicle in yaw–roll motion and the control objective of preventing the vehicle's rollover.

Part I

Tensor Product (TP) Model Formulation

Chapter 2

TP Model

This chapter defines the tensor product (TP) type model representation of quasi-linear parameter-varying (qLPV) state space models. Consider the following state space model:

$$\dot{\mathbf{x}}(t) = \mathbf{A}(\mathbf{p}(t))\mathbf{x}(t) + \mathbf{B}(\mathbf{p}(t))\mathbf{u}(t)$$
$$\mathbf{y}(t) = \mathbf{C}(\mathbf{p}(t))\mathbf{x}(t) + \mathbf{D}(\mathbf{p}(t))\mathbf{u}(t) \quad , \tag{2.1}$$

with input $\mathbf{u}(t) \in \mathbb{R}^k$, output $\mathbf{y}(t) \in \mathbb{R}^l$, and state vector $\mathbf{x}(t) \in \mathbb{R}^m$. The system matrix

$$\mathbf{S}(\mathbf{p}(t)) = \begin{pmatrix} \mathbf{A}(\mathbf{p}(t)) & \mathbf{B}(\mathbf{p}(t)) \\ \mathbf{C}(\mathbf{p}(t)) & \mathbf{D}(\mathbf{p}(t)) \end{pmatrix} \in \mathbb{R}^{(m+k)\times(m+l)} \tag{2.2}$$

is a parameter-varying quantity, where $\mathbf{p}(t) \in \Omega$ is a time-varying N-dimensional parameter vector within the closed hypercube $\Omega = [a_1, b_1] \times [a_2, b_2] \times \cdots \times [a_N, b_N] \subset \mathbb{R}^N$. If the parameter $\mathbf{p}(t)$ does not include any element of $\mathbf{x}(t)$, (2.1) is a linear parameter-varying (LPV) system. If the parameter $\mathbf{p}(t)$ does include some elements of $\mathbf{x}(t)$, system (2.1) then belongs to the class of nonlinear models. In this case, it is termed a qLPV model.

Note that throughout this book we consider the construction of TP model transformation for systems given in the form (2.1), but the same approach can be applied to construct TP model transformation for generalized multichannel systems of the form:

$$\begin{pmatrix} \dot{\mathbf{x}}(t) \\ \mathbf{z}_1(t) \\ \vdots \\ \mathbf{z}_q(t) \\ \mathbf{y}(t) \end{pmatrix} = \begin{pmatrix} \mathbf{A}(\mathbf{p}(t)) & \mathbf{B}_1(\mathbf{p}(t)) & \cdots & \mathbf{B}_q(\mathbf{p}(t)) & \mathbf{B}(\mathbf{p}(t)) \\ \mathbf{C}_1(\mathbf{p}(t)) & \mathbf{D}_{1,1}(\mathbf{p}(t)) & \cdots & \mathbf{D}_{1,q}(\mathbf{p}(t)) & \mathbf{E}_1(\mathbf{p}(t)) \\ \vdots & \vdots & \ddots & \vdots & \vdots \\ \mathbf{C}_q(\mathbf{p}(t)) & \mathbf{D}_{q,1}(\mathbf{p}(t)) & \cdots & \mathbf{D}_{q,q}(\mathbf{p}(t)) & \mathbf{E}_q(\mathbf{p}(t)) \\ \mathbf{C}(\mathbf{p}(t)) & \mathbf{F}_1(\mathbf{p}(t)) & \cdots & \mathbf{F}_q(\mathbf{p}(t)) & 0 \end{pmatrix} \begin{pmatrix} \mathbf{x}(t) \\ \mathbf{w}_1(t) \\ \vdots \\ \mathbf{w}_q(t) \\ \mathbf{u}(t) \end{pmatrix}, \tag{2.3}$$

where the elements of the system matrix of (2.3) may be dependent on $\mathbf{p}(t)$. Here, $\mathbf{w}_j(t) \to \mathbf{z}_j(t)$ are the channels on which we want to impose certain robustness and/or performance objectives. For simplicity, from now on we will drop the dependency on "t" from the variables if the meaning and context are clear. We will still keep "t" in some cases to facilitate better understanding.

Consider the case that the corresponding system matrix (2.2) is expressible as

$$\mathbf{S}(\mathbf{p}) = \sum_{h=1}^{H} w_h(\mathbf{p})\mathbf{S}_h \tag{2.4}$$

for any parameter vector $\mathbf{p}$, that is, $\mathbf{S}(\mathbf{p})$ can be given as a parameter-dependent linear combination of a set of linear time-invariant (LTI) system matrices, or vertex systems, $\mathbf{S}_h \in \mathcal{R}^{(m+k) \times (m+l)}$. System (2.1) then becomes

$$\begin{pmatrix} \dot{\mathbf{x}} \\ \mathbf{y} \end{pmatrix} = \left(\sum_{h=1}^{H} w_h(\mathbf{p})\mathbf{S}_h \right) \begin{pmatrix} \mathbf{x} \\ \mathbf{u} \end{pmatrix}. \tag{2.5}$$

Furthermore, let the multivariable weighting functions $w_h(\mathbf{p})$ in (2.4) and (2.5) also satisfy the conditions

$$w_h(\mathbf{p}) \in [0, 1] \tag{2.6}$$

and

$$\sum_{h=1}^{H} w_h(\mathbf{p}) = 1. \tag{2.7}$$

These two conditions imply that the linear combination in (2.4) to form $\mathbf{S}(\mathbf{p})$ is convex.

In this case, we can state that (2.5) is a finite element polytopic representation of system (2.1). For ease of terminology, we will still call (2.5) a model.

Definition 2.1 (Finite element polytopic model). System (2.5), with $w_h(\mathbf{p})$ satisfying (2.6) and (2.7), is a finite element polytopic model of system (2.1).

Finite element here means that the number H of the system matrices involved is bounded ($H < \infty$). The model is polytopic as the system matrix $\mathbf{S}(\mathbf{p})$ actually lies within a polytope defined in the parameter space Ω, with the linear time-invariant (LTI) systems $\mathbf{S}_h$ constituting the vertices of the polytope.

Before we define the finite element TP type polytopic model, or TP model for brevity, we introduce two basic operators of tensor algebra from the work of [DLDMV00] to facilitate compact notation for later discussion.

Definition 2.2 (n-mode matrix of tensor $\mathcal{A}$). Assume an Nth-order tensor $\mathcal{A} \in \mathbb{R}^{I_1 \times \cdots \times I_N}$. The n-mode matrix of $\mathcal{A}$, denoted as $\mathbf{A}_{(n)} \in \mathbb{R}^{I_n \times (I_{n+1} I_{n+2} \cdots I_N I_1 I_2 \cdots I_{n-1})}$, contains all the column vectors $\mathbf{a}_r \in \mathbb{R}^{I_n}$ of the nth dimension of tensor $\mathcal{A}$, where $r = 1, \ldots, R$ and $R = I_{n+1} I_{n+2} \cdots I_N I_1 I_2 \cdots I_{n-1}$. In general, the rth column of n-mode matrix $\mathbf{A}_{(n)}$ is equivalent to the $(i_1, i_2, \ldots, i_{n-1}, i_{n+1}, \ldots, i_N)$th vector of dimension I_n of $\mathcal{A}$, where $r = \text{ordering}(i_1, i_2, \ldots, i_{n-1}, i_{n+1}, \ldots, i_N)$. This *ordering* determines r as a linear index equivalent to the multilinear array index of the size $(I_1 \times I_2, \ldots, I_{n-1} \times I_{n+1}, \ldots, I_N)$. Note that this ordering of the column vectors can be arbitrarily determined. The important thing is that the same ordering and reordering must be adopted for all cases throughout.

Remark 2.1. For computational purposes, a possible choice for this ordering is:

$$
\begin{aligned}
r = \quad & (i_{n+1} - 1)I_{n+2}I_{n+3} \cdots I_N I_1 I_2 \cdots I_{n-1} + (i_{n+2} - 1)I_{n+3}I_{n+4} \cdots I_N I_1 I_2 \cdots I_{n-1} \\
& + \cdots + (i_N - 1)I_1 I_2 \cdots I_{n-1} + (i_1 - 1)I_2 I_3 \cdots I_{n-1} + (i_2 - 1)I_3 I_4 \cdots I_{n-1} \\
& + \cdots + (i_{n-2} - 1)I_{n-1} + i_{n-1}.
\end{aligned}
\tag{2.8}
$$

Equation (2.8) has $(N - 1)$ terms. As an example, Figure 2.1 shows the n-mode matrices, $\mathbf{A}_{(n)}$, $n = 1, 2, 3$, of a third-order tensor $\mathcal{A}$ formed in a way that is consistent with the index ordering r in Remark 2.1. Specifically, the top illustration shows how the $I_1 \times I_2 \times I_3$ tensor $\mathcal{A}$ is spread into different layers to form the $I_1 \times I_2 I_3$ matrix $\mathbf{A}_{(1)}$. The same spreading operation can be applied to form matrices $\mathbf{A}_{(2)}$ and $\mathbf{A}_{(3)}$ with the tensor $\mathcal{A}$ starting at different orientations of the indices.

In this example, where $N = 3$, we have $N - 1 = 2$ terms in (2.8), so

$$
r = (i_{n-2} - 1)I_{n-1} + i_{n-1}.
$$

For $n = 1$, $I_{n-1} = I_3$, $i_{n-1} = i_3$, $i_{n-2} = i_2$, the ordering for $\mathbf{A}_{(1)}$ is thus given by $r = (i_2 - 1)I_2 + i_3$. One can see that this is correct from Figure 2.1. The column of $\mathbf{A}_{(1)}$ is of dimension $I_1 \times 1$, and the rth column of $\mathbf{A}_{(1)}$ corresponding to the index i_1 and i_3 is indeed given by $r = (i_2 - 1)I_2 + i_3$. Similarly, the ordering for $\mathbf{A}_{(2)}$ is given by $r = (i_3 - 1)I_1 + i_1$, and that for the $\mathbf{A}_{(3)}$ matrix by $r = (i_1 - 1)I_1 + i_2$. The ordering (2.8) hence corresponds to the way the n-mode matrix is formed in Figure 2.1. Figure 2.2 shows another way to form the n-mode matrices $\mathbf{A}_{(n)}$ of the tensor $\mathcal{A}$, which would lead to a different ordering expression for r from (2.8). As the order of tensor $\mathcal{A}$ increases, the number of ways of forming the n-mode matrices also increases. In any case, as mentioned, once we decide upon a certain way to form the n-mode matrices and determine its corresponding ordering r, the key is to maintain the same throughout the whole process.

In the following we establish a notation for the multiplication of a higher-order tensor by a matrix. The notation will allow us to effectively express the transformation operations that we are going to apply later. Let us first look at the matrix product $\mathbf{G} = \mathbf{U} \cdot \mathbf{F} \cdot \mathbf{V}^T$ involving matrices $\mathbf{F} \in \mathbb{R}^{I_1 \times I_2}$, $\mathbf{U} \in \mathbb{R}^{J_1 \times I_1}$, $\mathbf{V} \in \mathbb{R}^{J_2 \times I_2}$, and $\mathbf{G} \in \mathbb{R}^{J_1 \times J_2}$. We observe that in the same way that the rows of $\mathbf{U}$ form linear combinations with the rows of $\mathbf{F}$, the rows of $\mathbf{V}$ (not $\mathbf{V}^T$) form linear combinations with the columns of

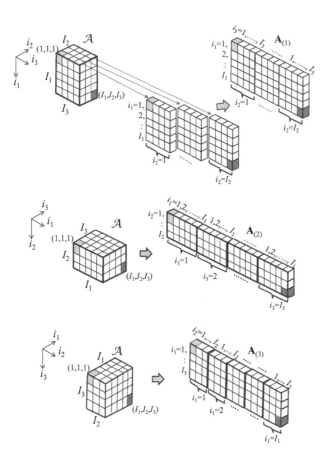

Figure 2.1: Formation of the n-mode matrices of a third-order tensor $\mathcal{A}$ with ordering given by (2.8).

F. Denoting the rows and columns of $\mathbf{F}$ as its first and second dimensions, we can express the relation $\mathbf{G} = \mathbf{U} \cdot \mathbf{F} \cdot \mathbf{V}^T$ by means of the $\times_n$ symbol: $\mathbf{G} = \mathbf{F} \times_1 \mathbf{U} \times_2 \mathbf{V}$. The new expression means that $\mathbf{F}$ is being multiplied along its first dimension (the rows) by $\mathbf{U}$, and along its second dimension (the columns) by $\mathbf{V}$. In general, we have the following definition:

Definition 2.3 (n-mode product of a tensor by a matrix). The n-mode product of a tensor $\mathcal{A} \in \mathbb{R}^{I_1 \times I_2 \times \cdots \times I_N}$ by a matrix $\mathbf{U} \in \mathbb{R}^{J_n \times I_n}$ along its nth dimension, denoted by $\mathcal{A} \times_n \mathbf{U}$, is an $(I_1 \times I_2 \times \cdots \times I_{n-1} \times J_n \times I_{n+1} \times \cdots \times I_N)$ tensor for which the entries

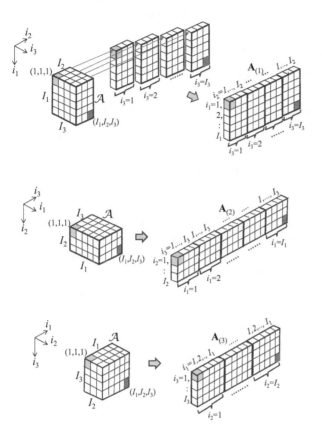

Figure 2.2: An alternative formation of the n-mode matrices of a third-order tensor $\mathcal{A}$.

are given by

$$(\mathcal{A} \times_n \mathbf{U})_{i_1,i_2,\ldots,i_{n-1},j_n,i_{n+1},\ldots,i_N} \overset{\text{def}}{=} \sum_{i_n} a_{i_1,i_2,\ldots,i_{n-1},i_n,i_{n+1},\ldots,i_N} u_{j_n,i_n}.$$

The multiple n-mode product of a tensor such as $\mathcal{A} \times_1 \mathbf{U}_1 \times_2 \mathbf{U}_2 \times_3 \cdots \times_N \mathbf{U}_N$ is denoted as

$$\mathcal{A} \times_1 \mathbf{U}_1 \times_2 \mathbf{U}_2 \times_3 \cdots \times_N \mathbf{U}_N \equiv \mathcal{A} \overset{N}{\underset{n=1}{\boxtimes}} \mathbf{U}_n,$$

which means that $\mathcal{A}$ is being multiplied along its nth dimension by $\mathbf{U}_n$ for $n = 1,\ldots,N$.

Remark 2.2. Computationally, the n-mode product of a tensor by a matrix, $\mathcal{A} = \mathcal{B} \times_n \mathbf{U}$, can be obtained by first finding the n-mode matrix of $\mathcal{B}$, $\mathbf{B}_{(n)}$, computing the product $\mathbf{A}_{(n)} = \mathbf{U}\mathbf{B}_{(n)}$, and then converting $\mathbf{A}_{(n)}$ to recover $\mathcal{A}$.

The n-mode product satisfies the following properties:

Property 2.1. Given the tensor $\mathcal{A} \in \mathbb{R}^{I_1 \times I_2 \times \cdots \times I_N}$ and the matrices $\mathbf{F} \in \mathbb{R}^{J_n \times I_n}$, $\mathbf{G} \in \mathbb{R}^{J_m \times I_m}$, $n \neq m$, we have

$$(\mathcal{A} \times_n \mathbf{F}) \times_m \mathbf{G} = (\mathcal{A} \times_m \mathbf{G}) \times_n \mathbf{F} = \mathcal{A} \times_n \mathbf{F} \times_m \mathbf{G}.$$

Property 2.2. Given the tensor $\mathcal{A} \in \mathbb{R}^{I_1 \times I_2 \times \cdots \times I_N}$ and the matrices $\mathbf{F} \in \mathbb{R}^{J_n \times I_n}$, $\mathbf{G} \in \mathbb{R}^{K_n \times J_n}$, we have

$$(\mathcal{A} \times_n \mathbf{F}) \times_n \mathbf{G} = \mathcal{A} \times_n (\mathbf{G} \cdot \mathbf{F}).$$

Now we return to the definition of the TP model. The TP model is a class of models in the form of (2.5), with the further property that the multivariable weighting functions $w_h(\mathbf{p})$ are decomposed as products of weighting functions of single variable p_n, $n = 1, \ldots, N$. This property leads to the incorporation of a tensor product structure in the ensuing system representation.

Definition 2.4 (Finite element TP type polytopic model, or TP model). A finite element polytopic model of the form (2.5), with $w_h(\mathbf{p}) = w_{1,i_1}(p_1) w_{2,i_2}(p_2) \ldots w_{N,i_N}(p_N)$ for any parameter $\mathbf{p}$, is a finite element TP type polytopic model, or TP model, of system (2.1).

The corresponding system matrix $\mathbf{S}(\mathbf{p})$ in (2.4) in this case becomes

$$\mathbf{S}(\mathbf{p}) = \sum_{i_1=1}^{I_1} \sum_{i_2=1}^{I_2} \cdots \sum_{i_N=1}^{I_N} \prod_{n=1}^{N} w_{n,i_n}(p_n) \mathbf{S}_{i_1,i_2,\ldots,i_N}, \tag{2.9}$$

for any parameter $\mathbf{p}$.

Comparing (2.9) with (2.4), we have $\sum_{h=1}^{H}$ replaced by $\sum_{i_1=1}^{I_1} \sum_{i_2=1}^{I_2} \cdots \sum_{i_N=1}^{I_N}$, $H = I_1 I_2 \cdots I_N$, $\mathbf{S}_h = \mathbf{S}_{i_1,i_2,\ldots,i_N}$, and $w_h(\mathbf{p}) = w_{1,i_1}(p_1) w_{2,i_2}(p_2) \ldots w_{N,i_N}(p_N)$. Here, h corresponds to a linear index that is equivalent to the multilinear array index of $i_1, i_2, \ldots, i_N$, similar to the ordering r defined as before in (2.8). One possible choice of $h = ordering(i_1, i_2, \ldots, i_N)$ is thus

$$h = (i_1 - 1) I_2 I_3 \cdots I_N + (i_2 - 1) I_3 \cdots I_N + \cdots + (i_{N-1} - 1) I_N + i_N. \tag{2.10}$$

Again, this ordering can be arbitrarily defined, except that the same ordering and reordering must be maintained during the process.

With the compact tensor notation introduced above, we can simply express (2.9) as:

$$\mathbf{S}(\mathbf{p}) = \mathcal{S} \underset{n=1}{\overset{N}{\boxtimes}} \mathbf{w}_n(p_n), \tag{2.11}$$

where the $(N + 2)$-dimensional coefficient tensor $\mathcal{S} \in \mathcal{R}^{I_1 \times I_2 \times \ldots I_N \times (m+k) \times (m+l)}$ is constructed from LTI vertex systems $\mathbf{S}_{i_1, i_2, \ldots, i_N} \in \mathcal{R}^{(m+k) \times (m+l)}$, and the row vector $\mathbf{w}_n(p_n) \in \mathcal{R}^{1 \times I_n}$ contains as elements the single variable continuous weighting functions $w_{n,i_n}(p_n) \in [0, 1]$, for $i_n = 1, \ldots, I_n$ and $n = 1, \ldots, N$. The function $w_{n,i_n}(p_n)$ is the i_nth weighting function defined on the nth dimension of Ω and p_n is the nth element of vector $\mathbf{p}$. The TP model is then:

$$\begin{pmatrix} \dot{\mathbf{x}} \\ \mathbf{y} \end{pmatrix} = \left(\mathcal{S} \underset{n=1}{\overset{N}{\boxtimes}} \mathbf{w}_n(p_n) \right) \begin{pmatrix} \mathbf{x} \\ \mathbf{u} \end{pmatrix}. \tag{2.12}$$

The TP model exhibits much more structure than the finite element polytopic model. Note that system (2.12) remains to be polytopic, that is, the weighting functions $w_h(\mathbf{p}) = w_{1,i_1}(p_1) w_{2,i_2}(p_2) \ldots w_{N,i_N}(p_N)$ still satisfy the conditions of (2.6) and (2.7) as finite element polytopic models.

To take advantage of the enhanced structure in the TP model, we desire a finer characterization of the polytopic conditions (2.6) and (2.7) in terms of $w_{n,i_n}(p_n)$, $n = 1, \ldots, N$, and $i_n = 1, \ldots, I_N$, instead of their product $w_h(\mathbf{p})$. We introduce the following two types of TP model.

Definition 2.5 (Nonnegativeness [NN] type TP model). The TP model (2.12) is NN type if its weighting functions satisfy for all $\mathbf{p} \in \Omega$,

$$\forall n, i, p_n : w_{n,i}(p_n) \geq 0. \tag{2.13}$$

Definition 2.6 (Sum normalization [SN] type TP model). The TP model (2.12) is SN (Sum Normalization) type if its weighting functions satisfy for all $\mathbf{p} \in \Omega$,

$$\forall n, p_n : \sum_{i_n=1}^{I_n} w_{n,i_n}(p_n) = 1. \tag{2.14}$$

Together, the NN and SN conditions imply that the weighting functions satisfy

$$\forall n, i, p_n : w_{n,i}(p_n) \in [0, 1], \tag{2.15}$$

$$\forall n, p_n : \sum_{i=1}^{I_n} w_{n,i}(p_n) = 1. \tag{2.16}$$

We note that condition (2.15) implies (2.6) straightforwardly, and that (2.16) implies (2.7) as:

$$\sum_{h=1}^{H} w_h(\mathbf{p}) = \sum_{i_1=1}^{I_1} \sum_{i_2=1}^{I_2} \ldots \sum_{i_N=1}^{I_N} \prod_{n=1}^{N} w_{n,i_n}(p_n) = \prod_{n=1}^{N} \left(\sum_{i_n=1}^{I_n} w_{n,i_n}(p_n) \right) = 1, \tag{2.17}$$

but the reverse is not true. This implies that the polytope defined by (2.15) and (2.16) is a tighter one over that defined by (2.6) and (2.7). The system matrix $\mathbf{S}(\mathbf{p})$ of an

NN and SN type TP model hence lies within a polytope more finely defined by the vertex LTI systems $S_{i_1,i_2,\ldots,i_N}$ for $\mathbf{p} \in \Omega$ of the tensor product structure.

From another perspective, it can be seen that conditions (2.15) and (2.16) refine the representation of the system matrix as a convex combination over multivariable coefficient functions of $\mathbf{p} \in \Omega$ to that over coefficient functions of single variable $p_n \in [a_n, b_n]$ for $n = 1, \ldots, N$. To this effect, we give a special characterization to the TP model satisfying both NN and SN conditions.

Definition 2.7 (Convex TP model). The TP model (2.12) is convex if it is both SN and NN type.

Here, convex means convex combination down to the p_n-variable function level. All mention of convexity in this book from now on will mean convexity at the level of the p_n-variable.

One example of a convex TP model is the popular Takagi–Sugeno (TS) fuzzy model of the form:

$$\text{IF } p_1 \in A_{1,i_1} \text{ and } p_2 \in A_{2,i_2} \text{ and } \cdots \text{ and } p_N \in A_{N,i_N}, \text{ THEN } S_{i_1,i_2,\ldots,i_N}, \quad (2.18)$$

for $n = 1, \ldots, N$, $i_n = 1, \ldots, I_n$. In this case, the fuzzy variables p_n, $n = 1, \ldots, N$, constitute the parameter $\mathbf{p}$, the membership functions of the antecedent A_{n,i_n} of fuzzy variable p_n constitute the weighting functions $w_{n,i_n}(p_n)$, and the consequents of the fuzzy rules constitute the LTI matrix elements $S_{i_1,i_2,\ldots,i_N}$. By adopting the Ruspini-partition of the antecedent fuzzy sets, the output of the TS model (2.18) under Product-Sum-Gravity inference is given by (2.9) satisfying the SN and NN conditions.

It will be shown in Chapter 6 that one can always incorporate the SN and NN conditions to a TP model through our techniques. Hence, a TP model is always a convex TP model. The SN and NN types are sort of the basic convexities for TP models. Besides SN and NN, there are other convexity conditions, which may or may not be possibly incorporated for a given TP model. The definition and incorporation of these convexity types will be given in Chapter 6. Incorporating different convexity conditions into the TP model amounts to varying the shape of the polytope, or convex hull, enclosing the system matrix. This ability makes way for numerous advantages in the linear matrix inequality (LMI)-based control system design.

As a final note in this chapter, we define the exact and nonexact TP models in the approximation of the original qLPV system matrix.

Definition 2.8 (Exact/Nonexact TP model). A TP model is termed an Exact TP model if for all $\mathbf{p} \in \Omega$,

$$S(\mathbf{p}) = S \underset{n=1}{\overset{N}{\boxtimes}} \mathbf{w}_n(p_n) \tag{2.19}$$

holds. A TP model is termed a Nonexact TP model if we can only find

$$\hat{\mathbf{S}}(\mathbf{p}) = \mathcal{S} \underset{n=1}{\overset{N}{\boxtimes}} \mathbf{w}_n(p_n) \tag{2.20}$$

as an approximation to $\mathbf{S}(\mathbf{p})$. In this case, the approximation error γ is defined as:

$$\max_{\mathbf{p}} \|\mathbf{S}(\mathbf{p}) - \hat{\mathbf{S}}(\mathbf{p})\|_{\mathcal{L}_2} = \gamma. \tag{2.21}$$

Chapter 3

TP Model Transformation

The key idea of tensor product (TP) model transformation was first introduced in [BTYP03], [Bar04], [BVK05]. The objective of the transformation is to numerically convert any given quasi-linear parameter-varying (qLPV) model (2.1) into a finite element TP type polytopic model type form (Definition 2.4) in the parameter space Ω:

$$\begin{pmatrix} \dot{\mathbf{x}} \\ \mathbf{y} \end{pmatrix} = \mathbf{S}(\mathbf{p}) \begin{pmatrix} \mathbf{x} \\ \mathbf{u} \end{pmatrix} = \left(S \underset{n=1}{\overset{N}{\boxtimes}} \mathbf{w}_n(p_n) \right) \begin{pmatrix} \mathbf{x} \\ \mathbf{u} \end{pmatrix}. \tag{3.1}$$

Particularly, the TP model transformation generates a canonical form as an intermediate step during the process. This canonical form, to be discussed in detail in Chapter 4, will enable the incorporation and manipulation of different types of TP model convexity as defined in Chapters 2 and 6. Given that the ensuing controller design feasibility and performance are very sensitive to the type of convexity attained, the ability to carry out such manipulations in the design process is highly desirable. The transformation also allows the determination of exact and nonexact TP model representations (Definition 2.8) for the given qLPV model using only a minimal number of components. In the nonexact case, a trade-off study between the number of components and the accuracy of the resulting TP model is also provided.

The procedures of the TP model transformation involve the discretization of the given qLPV model, and then using higher-order singular value decomposition (HOSVD) to obtain the unique tensor product structure of the given model. Based on this tensor structure we can readily identify the various components of the resultant TP model. *Overall, the TP model transformation can be viewed as a HOSVD of the qLPV models which also serves as a means of convex hull manipulation for control system design.*

21

3.1 Introduction to HOSVD

We will first briefly introduce the HOSVD of tensors that forms the basis of TP model transformation. The precursor of the HOSVD technique can be found in [Yam97], where the method was applied for fuzzy rule-based approximation of multivariable functions. Soon after, a comprehensive analysis of singular value decomposition (SVD) application to Nth-order tensors appears in [DLDMV00]. In the following presentation, we will adopt the notations in Lathauwer's work due to their compact forms.

We recall the results on matrix SVD as follows:

Theorem 3.1 (Matrix SVD). *Every real $(I_1 \times I_2)$ matrix $\mathbf{A}$ can be written as the product*

$$\mathbf{A} = \mathbf{U}_1 \cdot \mathbf{S} \cdot \mathbf{U}_2^T = \mathbf{S} \times_1 \mathbf{U}_1 \times_2 \mathbf{U}_2 = \mathbf{S} \overset{2}{\underset{n=1}{\boxtimes}} \mathbf{U}_n, \tag{3.2}$$

in which

1. *$\mathbf{U}_1 = (\mathbf{u}_{1,1}\mathbf{u}_{1,2}\cdots\mathbf{u}_{1,I_1})$ is an orthonormal $(I_1 \times I_1)$ matrix,*

2. *$\mathbf{U}_2 = (\mathbf{u}_{2,1}\mathbf{u}_{2,2}\cdots\mathbf{u}_{2,I_2})$ is an orthonormal $(I_2 \times I_2)$ matrix,*

3. *$\mathbf{S}$ is an $(I_1 \times I_2)$ matrix with the properties of*

   (a) pseudodiagonality:

   $$\mathbf{S} = \mathrm{diag}\left(\sigma_1, \sigma_2, \ldots, \sigma_{\min(I_1,I_2)}\right), \tag{3.3}$$

   (b) ordering:

   $$\sigma_1 \geq \sigma_2 \geq \cdots \geq \sigma_{\min(I_1,I_2)} \geq 0. \tag{3.4}$$

The σ_i are the singular values of $\mathbf{A}$ and the vectors $\mathbf{u}_{1,i}$ and $\mathbf{u}_{2,i}$ are, respectively, an ith left and an ith right singular vector.

Remark 3.1. The number of nonzero singular values σ_i is equal to the rank of matrix $\mathbf{A}$.

We also extend the well-known definitions of scalar product, orthogonality, and Frobenius-norm to tensors as:

Definition 3.1 (Scalar product). The scalar product $\langle \mathcal{A}, \mathcal{B} \rangle$ of two tensors $\mathcal{A}, \mathcal{B} \in \mathbb{R}^{I_1 \times I_2 \times \cdots \times I_N}$ is defined as

$$\langle \mathcal{A}, \mathcal{B} \rangle \overset{\text{def}}{=} \sum_{i_1=1}^{I_1} \sum_{i_2=1}^{I_2} \cdots \sum_{i_N=1}^{I_N} b_{i_1,i_2,\ldots,i_N} a_{i_1,i_2,\ldots,i_N}.$$

Definition 3.2 (Orthogonality). Tensors of which the scalar product equals 0 are orthogonal.

Definition 3.3 (Frobenius-norm). The Frobenius-norm of a tensor $\mathcal{A}$ is given by

$$\|\mathcal{A}\| \overset{\text{def}}{=} \sqrt{\langle \mathcal{A}, \mathcal{A} \rangle}.$$

We now define the n-mode rank of tensors. We take notice that there are major differences between matrices and higher-order tensors as far as their rank properties are concerned. Moreover, there are many theories in the literature regarding the rank properties of tensors themselves. Here we adopt the definition that constitutes a simple generalization of the notion of column and row rank:

Definition 3.4 (n-mode rank of tensor $\mathcal{A}$). The n-mode rank of tensor $\mathcal{A}$, denoted by $R_n = \text{rank}_n(\mathcal{A})$, is the dimension of the vector space spanned by the n-mode matrix of tensor $\mathcal{A}$.

Remark 3.2. Computationally, the n-mode rank, $\text{rank}_n(\mathcal{A})$, can be readily obtained by the rank of the n-mode matrix $\mathbf{A}_{(n)}$, that is, $\text{rank}_n(\mathcal{A}) = \text{rank}\,(\mathbf{A}_{(n)})$.

Now we state the generalization of matrix SVD to tensors:

Theorem 3.2 (Higher-order SVD [HOSVD]). *Every real* $(I_1 \times I_2 \times \cdots \times I_N)$ *tensor* $\mathcal{A}$ *can be written as the product*

$$\mathcal{A} = S \times_1 \mathbf{U}_1 \times_2 \mathbf{U}_2 \times_3 \cdots \times_N \mathbf{U}_N = S \overset{N}{\underset{n=1}{\boxtimes}} \mathbf{U}_n, \tag{3.5}$$

in which

1. $\mathbf{U}_n = (\mathbf{u}_{n,1}\mathbf{u}_{n,2}\cdots\mathbf{u}_{n,I_n})$, $n = 1, \ldots, N$ *is an orthonormal* $(I_n \times I_n)$ *matrix,*

2. S *is a real* $(I_1 \times I_2 \times \cdots \times I_N)$ *tensor of which the subtensors* $S_{i_n=\alpha}$, *obtained by fixing the nth index to* α, *have the properties of*

    (a) *all-orthogonality: two subtensors* $S_{i_n=\alpha}$ *and* $S_{i_n=\beta}$ *are orthogonal for all possible values of* n, α, *and* β *subject to* $\alpha \neq \beta$:

    $$\langle S_{i_n=\alpha}, S_{i_n=\beta} \rangle = 0, \text{ when } \alpha \neq \beta, \tag{3.6}$$

    (b) *ordering:*
    $$\|S_{i_n=1}\| \geq \|S_{i_n=2}\| \geq \cdots \geq \|S_{i_n=I_n}\| \geq 0, \tag{3.7}$$
    for all possible values of n.

The Frobenius-norms $\|S_{i_n=i}\|$, *symbolized by* $\sigma_i^{(n)}$, *are n-mode singular values of* $\mathcal{A}$, *the vectors* $\mathbf{u}_{n,i}$ *are the ith n-mode singular vectors, and the matrix* $\mathbf{U}_n$ *is the n-mode singular matrix.*

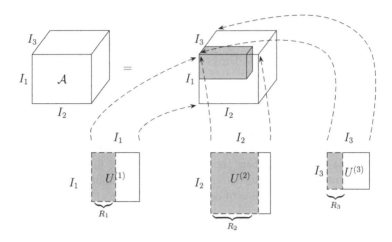

Figure 3.1: Visualization of the HOSVD for a third-order tensor.

Figure 3.1 presents an illustration of the HOSVD operation for a third-order tensor. Note that the HOSVD uniquely determines tensor S, but the determination of matrices $\mathbf{U}_n$ may not be unique if there are equivalent singular values in at least one of the dimensions.

Remark 3.3. The n-mode singular matrix $\mathbf{U}_n$ (and the n-mode singular values) can be directly found as the left singular matrix (and the singular values) of an n-mode matrix of tensor $\mathcal{A}$. Hence computing the HOSVD of an Nth-order tensor leads to the computation of N different matrix SVDs of matrices with size $(I_n \times I_1 I_2 \ldots I_{n-1} I_{n+1} \ldots I_N)$:

$$\mathbf{A}_{(n)} = \mathbf{U}_n \Theta_n \mathbf{V}_n^T$$

for $n = 1, \ldots, N$. Afterwards, the core tensor S can be computed by bringing the matrices of singular vectors to the left side of (3.5):

$$S = \mathcal{A} \times_1 \mathbf{U}_1^T \times_2 \mathbf{U}_2^T \times_3 \cdots \times_N \mathbf{U}_N^T = \mathcal{A} \underset{n=1}{\overset{N}{\boxtimes}} \mathbf{U}_n^T,$$

and using Property 2.2.

The minimal or compact form of HOSVD is defined as:

Definition 3.5 (Compact higher-order SVD [CHOSVD])**.** The HOSVD is computed by executing SVD for each n-mode matrix of $\mathcal{A}$, $n = 1, \ldots, N$. If we discard the zero singular values and the related singular vectors $\mathbf{u}_{R_n+1,n}, \mathbf{u}_{R_n+2,n}, \ldots, \mathbf{u}_{I_N,n}$, where $R_n = \text{rank}_n(\mathcal{A})$, during the SVD computation at each dimension, then we obtain the minimized form of HOSVD as

$$\mathcal{A} = \mathcal{D} \underset{n=1}{\overset{N}{\boxtimes}} \mathbf{U}_n, \tag{3.8}$$

which has all the properties as in Theorem 3.2 except that the sizes of $\mathbf{U}_n$ and $\mathcal{D}$ are $(I_n \times R_n)$ and $(R_1 \times \cdots \times R_N)$, respectively, with $R_n \leq I_n$.

Figure 3.1 depicts the situation in the CHOSVD of a third-order tensor, where R_1, R_2, and R_3 singular values and corresponding singular vectors (shown in grey) are kept in the respective dimensions.

We call the form (3.8) compact as $\mathcal{A}$ still retains its original values after discarding the zero singular values, except that the sizes of $\mathbf{U}_n$ and $\mathcal{D}$ are reduced to their minimal possible values. Should we go further and discard some of the nonzero singular values, we obtain the reduced form of HOSVD.

Definition 3.6 (Reduced higher-order SVD [RHOSVD]). If we discard some nonzero singular values and their corresponding singular vectors in HOSVD, the result is an approximation of tensor $\mathcal{A}$ with the following property.

Property 3.1. Assume the HOSVD of tensor $\mathcal{A}$ is given according to Theorem 3.2 as $\mathcal{A} = \mathcal{S} \overset{N}{\underset{n=1}{\boxtimes}} \mathbf{U}_n$, and the n-mode rank of $\mathcal{A}$ is R_n, $n = 1, \ldots, N$. If we discard the nonzero singular values $\sigma_{R'_n+1}^{(n)}, \sigma_{R'_n+2}^{(n)}, \ldots, \sigma_{R_n}^{(n)}$ and the corresponding vectors of the singular matrices during the SVD of the n-mode matrix of $\mathcal{S}$ for a chosen $R'_n < R_n$, then we obtain

$$\hat{\mathcal{A}} = \mathcal{D} \overset{N}{\underset{n=1}{\boxtimes}} \mathbf{U}_n, \tag{3.9}$$

where the sizes of $\mathbf{U}_n$ and $\mathcal{D}$ are $(I_n \times R'_n)$ and $(R'_1 \times \cdots \times R'_N)$, respectively. In this case, the difference between $\mathcal{A}$ and $\hat{\mathcal{A}}$ is characterized by

$$\gamma = \|\mathcal{A} - \hat{\mathcal{A}}\|^2 \leq \sum_{i_1=R'_1+1}^{R_1} \left(\sigma_{i_1}^{(1)}\right)^2 + \sum_{i_2=R'_2+1}^{R_2} \left(\sigma_{i_2}^{(2)}\right)^2 + \cdots + \sum_{i_N=R'_N+1}^{R_N} \left(\sigma_{i_N}^{(N)}\right)^2. \tag{3.10}$$

RHOSVD will be used later when we trade off the number of singular values to keep and the would-be approximation error for the nonexact TP model representation of a given qLPV model. This trade-off between complexity and accuracy of the ensuing TP model will have a great impact on the linear matrix inequality (LMI) controller design step to come. Property 3.1 is the Nth-order generalization of the relation between the singular value decomposition of a matrix and its best, lower-ranked matrix approximation (in least square sense). Specifically, for the matrix case, $N = 2$, discarding the smallest singular value from $\mathcal{A}$ would result in a matrix $\hat{\mathcal{A}}$ with rank number lowered by 1 in both dimensions, and that $\hat{\mathcal{A}}$ is then the best approximation (i.e., lowest γ) to $\mathcal{A}$ for all possible matrices with the same rank condition. In the higher-order case, $N \geq 2$, however, this is not true. Discarding the smallest singular values in any one dimension does result in a lower rank along all dimensions, but the resultant tensor $\hat{\mathcal{A}}$ may not necessarily be the best approximation (i.e., lowest γ) to the given tensor $\mathcal{A}$ for the same rank condition. Nevertheless, the ordering of the magnitudes of the singular values does provide an indication of the resulting error should they be discarded, and the upper limit of this error is given by (3.10). It turns out that the best approximation for a given rank reduction can be achieved by the proper modification of the elements of tensor $\hat{\mathcal{A}}$. For further details see higher-order orthogonal iteration (HOOI) in [DLDMV97], [IDLAVH08].

3.2   Transformation procedures

We present the procedures to transform a given qLPV model into a TP model form.

Step 1: Discretization

This step is to numerically convert the given parameter-dependent system matrix $\mathbf{S(p)}$ into a tensor representation to facilitate tensor product structure extraction. The transformation space Ω in which we expect the system to be valid is first defined.

Definition 3.7 (Transformation space Ω). Ω is a bounded hyper-rectangular space where the parameter vector of the system matrix varies: $\mathbf{p} \in \Omega : [a_1, b_1] \times [a_2, b_2] \times \cdots \times [a_N, b_N]$. Practically, Ω should be defined according to the working space of $\mathbf{p}$ determined based on the physical consideration of the model. We should note that the weighting functions of the resulting TP model will be defined over the intervals $[a_n, b_n], n = 1, \ldots, N$, so the resulting TP model is interpretable only in Ω.

Definition 3.8 (Discretization grid M). M denotes a hyper-rectangular discretization grid defined in Ω. M_n denotes the number of grid lines in the nth dimension for $n = 1, \ldots, N$, and $a_n \leq g_{n,m_n} \leq b_n, m_n = 1, \ldots, M_n$, denote the corresponding locations of the grid lines. Generally, the grid lines can be arbitrarily located in the intervals; however, we suggest defining an equidistant location: $g_{n,m_n} = a_n + \frac{b_n - a_n}{M_n - 1}(m_n - 1)$, if there is no special reason for doing otherwise. Moreover, our experience suggests defining a prime number M_n of grid lines in all dimensions. A general grid point in Ω is thus defined by the coordinate vector

$$\mathbf{g}_{m_1, m_2, \ldots, m_N} = \begin{pmatrix} g_{1,m_1} \\ \vdots \\ g_{N,m_N} \end{pmatrix}. \tag{3.11}$$

Definition 3.9 (Discretization of the qLPV model). $\mathcal{S}^D$ (superscript "D" denotes "discretized") denotes the discretized system matrix $\mathbf{S(p)} \in \mathbb{R}^{(m+k) \times (m+l)}$ over the hyper-rectangular grid M in Ω. This means that the entries of $\mathcal{S}^D$ are

$$\mathcal{S}_{m_1, m_2, \ldots, m_N} = \mathbf{S}(\mathbf{g}_{m_1, m_2, \ldots, m_N}), \tag{3.12}$$

for $m_n = 1, \ldots, M_n, n = 1, \ldots, N$. Thus, the size of $\mathcal{S}^D$ is $(M_1 \times M_2 \times \cdots \times M_N \times (m + k) \times (m + l))$. Note that we can define as dense a discretization grid as needed to guarantee that the discretized tensor describes the system matrix with sufficient precision in Ω. Specifically, we desire to express S^D according to the form (2.11):

$$S^D = S \underset{n=1}{\overset{N}{\boxtimes}} \mathbf{w}_n^D(p_n),\tag{3.13}$$

where $\mathbf{w}_n^D(p_n)$ are the values of the weighting functions $\mathbf{w}_n(p_n)$ evaluated at the discretized values of $p_n = g_{n,m_n}$, $m_n = 1, \ldots, M_n$ over the n-dimension interval $[a_n, b_n]$.

Step 2: Application of HOSVD

This step calls for executing HOSVD on the discretized system tensor S^D of Step 1. It is important to note that execution is on the first N dimension only, after which we have:

$$S^D = S \underset{n=1}{\overset{N}{\boxtimes}} \mathbf{U}_n.\tag{3.14}$$

Here, we denote the number of singular values and vectors retained at each dimension as I_n, and hence the size of S is $(I_1 \times I_2 \times \ldots \times I_N \times (m + k) \times (m + l))$ and that of $\mathbf{U}_n$ as $(M_n \times I_n)$. If we are executing CHOSVD, then $I_n = R_n = rank_n(S^D)$. If we are executing RHOSVD, then $I_n < R_n$ at least for one n. Hence, normally we have $I_n \le M_n$ for all $n = 1, \ldots, N$. The tensor S contains the $((m + k) \times (m + l))$ matrices $\mathbf{S}_{i_1, i_2, \ldots, i_N}$, $i_n = 1, \ldots, I_n$ as elements.

One may further define matrix transformations $\mathbf{T}_n$ to transform singular matrices $\mathbf{U}_n$ to $\bar{\mathbf{U}}_n$:

$$\bar{\mathbf{U}}_n' \mathbf{T}_n = \mathbf{U}_n.\tag{3.15}$$

In this case, (3.14) can be expressed as:

$$S^D = S \underset{n=1}{\overset{N}{\boxtimes}} \mathbf{U}_n = S \underset{n=1}{\overset{N}{\boxtimes}} \left(\bar{\mathbf{U}}_n \mathbf{T}_n \right) = S \underset{n=1}{\overset{N}{\boxtimes}} \mathbf{T}_n \underset{n=1}{\overset{N}{\boxtimes}} \bar{\mathbf{U}}_n = \bar{S} \underset{n=1}{\overset{N}{\boxtimes}} \bar{\mathbf{U}}_n,\tag{3.16}$$

where

$$\bar{S} = S \underset{n=1}{\overset{N}{\boxtimes}} \mathbf{T}_n.\tag{3.17}$$

If transformation $\mathbf{T}_n$ is defined in a special way, we can incorporate into $\bar{\mathbf{U}}_n$ certain convexity types, and hence desirable convex hull, for the later control system design process. Such transformations will be discussed in Chapter 6. For now, assume that we have (3.14) after Step 2.

Step 3: Identification of components

At the end of step 2, we have the form (3.14), where $\mathbf{U}_n$ may be obtained from CHOSVD, RHOSVD, or as a result of special transformation $\mathbf{T}_n$. Comparing (3.14) with (3.13), we can readily identify

- I_n as the number of weighting functions along the n-dimension, for $n = 1, \ldots, N$.

- The $((m + k) \times (m + l))$ matrices $\mathbf{S}_{i_1, i_2, \ldots, i_N}$, $i_n = 1, \ldots, I_n$ as the linear time-invariant (LTI) systems.

- The matrix $\mathbf{U}_n \in \mathbb{R}^{M_n \times I_n}$ as $\mathbf{w}_n^D(p_n)$, that is, the nth dimension weighting functions evaluated as the discretized p_n values over the domain $[a_n, b_n]$. Specifically, the i_nth column vector $\mathbf{u}_{n,i_n}$ of the matrix $\mathbf{U}_n$ determines the i_nth column vector $w_{n,i_n}(p_n)$ of $\mathbf{w}_n^D(p_n)$ discretized over $a_n \leq g_{n,m_n} \leq b_n$ for $m_n = 1, \ldots, M_n$. Thus, the m_nth element of $\mathbf{u}_{n,i_n}$, denoted as $\mathbf{u}_{n,i_n}(m_n)$, defines the values of the i_nth weighting function $w_{n,i_n}(p_n)$ at the point $p_n = g_{n,m_n}$:

$$w_{n,i_n}(g_{n,m_n}) = u_{n,i_n}(m_n).$$

The weighting functions above are obtained only at the discretized grid lines g_{n,m_n} in the nth dimension interval $[a_n, b_n]$. In order to generate continuous weighting functions, as those in (2.11), one can interpolate the discretized values to form continuous functions over $[a_n, b_n]$. In this book, we adopt linear interpolation for simplicity. It will be seen later that such interpolation would yield an acceptable approximation error covered by the robustness of the resultant controller. Moreover, linear interpolation also has the advantage that whatever convexity condition incorporated later at the discretized points through the transformation $\mathbf{T}_n$ in (3.15) will also be maintained at the in-between points.

3.3   The extracted model

As a result, the TP model transformation yields the following numerical model structure for the qLPV model $\mathbf{S}(\mathbf{p})$:

$$\tilde{\mathbf{S}}(\mathbf{p}) = \mathcal{S} \underset{n=1}{\overset{N}{\boxtimes}} \tilde{\mathbf{U}}_n, \tag{3.18}$$

where $\tilde{\mathbf{U}}_n$ denotes the set of continuous weighting functions formed by linear interpolating elements of the column vector $\mathbf{u}_{n,i_n}$ of $\mathbf{U}_n$ of (3.14) over the nth dimension interval $[a_n, b_n]$. Hence, we are adopting a piecewise linear approximation of $\mathbf{S}(\mathbf{p})$ using $\tilde{\mathbf{S}}(\mathbf{p})$. If CHOSVD is executed, then the value of $\mathbf{S}(\mathbf{p})$ and $\tilde{\mathbf{S}}(\mathbf{p})$ are the same at the grid points. If RHOSVD is executed, then the difference between $\mathbf{S}(\mathbf{p})$ and $\tilde{\mathbf{S}}(\mathbf{p})$ at the grid points would be characterized by (3.10).

Note that the extracted model (3.18) is based on (3.14) prior to the incorporation of any convexity condition. It is generally not polytopic while exhibiting a TP structure. The HOSVD process normally does not produce a $\mathbf{U}_n$ satisfying the polytopic conditions. The linearly interpolated continuous version $\tilde{\mathbf{U}}_n$ will hence also not be polytopic. As is, the model (3.18) is not usable for our control system design. Nevertheless, it does possess some interesting properties. The model actually constitutes the HOSVD-based TP canonical model, which will be discussed in the next chapter.

Furthermore, the complexity of the model (3.18) is dictated by the sizes I_n, $n = 1, \ldots, N$. Specifically, the number of LTI systems is given by $I_1 I_2 \ldots I_N$. Should we desire to decrease the number of LTI systems, we can execute RHOSVD in Step 2 above and discard more nonzero singular values and vectors to yield reduced sizes of I_n. Again, the maximum error γ between $\mathbf{S}(\mathbf{p})$ and $\tilde{\mathbf{S}}(\mathbf{p})$ at the grid points would be reflected by (3.10). The present process hence provides a means of trading off between the TP model complexity and approximation error by the adjustment of the number of discarded nonzero singular values and vectors in conducting RHOSVD. This trade-off process will be the subject of Chapter 5.

The incorporation of convexity conditions to the extracted model will be given in Chapter 6. Upon incorporation, model (3.18) will become a TP model ready to be used for control system design. In the ensuing LMI-based control method, particularly the parallel distributed compensation (PDC) framework, that we are going to apply in the book, only the LTI systems in S of (3.18) will be used for design. The weighting functions $\tilde{\mathbf{U}}_n$ will not be involved in the design process except to ensure that the model (3.18) is convex. Only after we have attained the LTI feedback from the LTI systems will the weighting functions play a role in combining the feedback to form the overall controller. With this view, it can be stated that the priority of the TP model transformation would be to produce a sufficiently precise computation of the LTI systems. And we rely on the robustness of the designed controller to compensate for using the linearly interpolated approximations of the weighting functions.

Remark 3.4. The maximum error of the extracted TP model at the grid points characterized by (3.10) can actually be extended to the whole domain of Ω. This is because we are using linear interpolation and the error in between grid points must be bounded by those at the grid points. Equation (3.10) is, however, a highly conservative estimate. A better characterization of the error between the resulting TP model and the original model can be obtained by numerically sampling the values over a huge number of points in Ω.

3.4 Addition of sampling grid lines

The weighting functions $w_{n,i_n}(p_n)$ determined above are at the discretized points g_{n,m_n} of the interval $[a_n, b_n]$. Consider the case where, for want of a denser sampling, say, we add one more grid line at $p_d = g_d^a$ in the interval $[a_d, b_d]$ along the d-dimension, thereby increasing the number of grid lines within $[a_d, b_d]$ from the original M_d to $M_d + 1$. We desire the values of the d-dimension weighting functions at the new grid point $p_d = g_d^a$. One way to proceed, of course, would be to discretize the system matrix $\mathbf{S}(\mathbf{p}) \in \mathbb{R}^{(m+k) \times (m+l)}$ over the expanded hyper-rectangular grid, which includes the additional grid line $p_d = g_d^a$ in the d-dimension, and apply HOSVD to yield the values of d-dimension weighting functions at the new grid line. This is, however, heavily computation-intensive. We can actually obtain the information using the TP model structure already extracted from $\mathcal{S}^D$ without the need of applying HOSVD anew.

We discretize the the system matrix $\mathbf{S}(\mathbf{p}) \in \mathbb{R}^{(m+k) \times (m+l)}$ over the single grid point $\bar{\mathbf{p}}$ in the expanded hyper-rectangular grid, with

$$
\bar{\mathbf{p}} = \begin{pmatrix} g_{1,\bar{m}_1} \\ \vdots \\ g_{d-1,\bar{m}_{d-1}} \\ g_d^a \\ g_{d+1,\bar{m}_{d+1}} \\ \vdots \\ g_{N,\bar{m}_N} \end{pmatrix}, \tag{3.19}
$$

where $\bar{m}_n$, $n \neq d$, is any designated index selected from $[1, \ldots, M_n]$. Apart from $p_d = g_d^a$ in the d-dimension, the value $p_n = g_{n,\bar{m}_n}$ corresponds to one of the original grid lines in the n-dimension, $n \neq d$. Then, referring to (3.13), we have

$$
\mathbf{S}(\bar{\mathbf{p}}) = \left(\mathcal{S} \times_d \mathbf{w}_d^D(g_d^a) \right) \underset{\substack{n=1 \\ n \neq d}}{\overset{N}{\boxtimes}} \mathbf{w}_n^D(g_{n,\bar{m}_n}), \tag{3.20}
$$

where $\mathbf{w}_d^D(g_d^a) \in \mathbb{R}^{1 \times I_d}$ is the desired d-dimension weighting functions evaluated at the new grid line $p_d = g_d^a$ and $\mathbf{w}_n^D(g_{n,\bar{m}_n}) \in \mathbb{R}^{1 \times I_n}$ is the n-dimension weighting functions evaluated at the designated original grid line $p_n = g_{n,\bar{m}_n}$. With the fact that $\mathbf{w}_n^D(g_{n,\bar{m}_n})$ is the $\bar{m}_n$th row of $\mathbf{U}_n$, we denote $\mathbf{w}_n^D(g_{n,\bar{m}_n})$ as $\mathbf{u}_n^{(\bar{m}_n)}$ and rewrite $\mathbf{S}(\bar{\mathbf{p}})$ as

$$
\mathbf{S}(\bar{\mathbf{p}}) = \left(\mathcal{S} \times_d \mathbf{w}_d^D(g_d^a) \right) \underset{\substack{n=1 \\ n \neq d}}{\overset{N}{\boxtimes}} \mathbf{u}_n^{(\bar{m}_n)}. \tag{3.21}
$$

Equation (3.21) constitutes a linear equation to solve for the unknown $\mathbf{w}_d^D(g_d^a)$. One should note that quantities on both sides of the equation are $(1 \times 1 \times \cdots \times 1 \times (m + k) \times (m + l))$ tensors. Taking the d-mode matrices of the tensors on both sides yields

$$
(\mathbf{S}(\bar{\mathbf{p}}))_{(d)} = \mathbf{w}_d^D(g_d^a) \left(\mathcal{S} \underset{\substack{n=1 \\ n \neq d}}{\overset{N}{\boxtimes}} \mathbf{u}_n^{(\bar{m}_n)} \right)_{(d)}. \tag{3.22}
$$

This gives the solution of

$$
\mathbf{w}_d^D(g_d^a) = (\mathbf{S}(\bar{\mathbf{p}}))_{(d)} \left[\left(\mathcal{S} \underset{\substack{n=1 \\ n \neq d}}{\overset{N}{\boxtimes}} \mathbf{u}_n^{(\bar{m}_n)} \right)_{(d)} \right]^+, \tag{3.23}
$$

where the superscript "+" denotes pseudo-inverse. Equation (3.23) can usually be solved but it may also be underdetermined or numerically ill-conditioned. In that case, we can generate more equations to solve for $\mathbf{w}_d^D(g_d^a)$ by discretizing the system matrix $\mathbf{S}(\mathbf{p})$ at different $\bar{\mathbf{p}}$ points through the picking of different sets of designated indices $\bar{m}_n$.

Note that it would be difficult to tell in advance the designated indices and hence grid points to pick in $\bar{\mathbf{p}}$ to effectively result in a solvable equation, but the number of useful equations can surely be increased by incrementally picking more $\bar{\mathbf{p}}$ until a solution is attainable. The above formulation deals with the case that we add just one grid line in the d-dimension. The same applies if we were to add multiple grid lines.

Remark 3.5. It is important to emphasize here that the discretization density of grid M is restricted by the heavy computational load of the HOSVD. However, in the present formulation we do not need to calculate HOSVD and we can readily determine the discretized weighting functions over $L_n \gg M_n$ points in any dimension if better approximation of the weighting functions is needed.

Remark 3.6. *The TP model transformation can be viewed as computationally equivalent to an analytical expression of the qLPV model. It is capable of producing the LTI systems as well as the weighting functions over any points, as if with an analytically derived closed form expression.*

Chapter 4

TP Canonical Model Form

This chapter defines the tensor product (TP) canonical form for quasi-linear parameter-varying (qLPV) models. Furthermore, it shows that the TP model transformation of Chapter 3 is capable of numerically reconstructing such canonical form in the extract model (3.18). The original idea of the chapter can be found in [BSVY06].

4.1 Definition

Consider a given TP model

$$
\begin{pmatrix} \dot{\mathbf{x}} \\ \mathbf{y} \end{pmatrix} = \left(S \underset{n=1}{\overset{N}{\boxtimes}} \mathbf{w}_n(p_n) \right) \begin{pmatrix} \mathbf{x} \\ \mathbf{u} \end{pmatrix}, \tag{4.1}
$$

where $\mathbf{p} \in \Omega$ is the time-varying N-dimensional parameter vector and S is $(I_1 \times I_2 \times \cdots \times I_N \times I_{N+1} \times I_{N+2})$ (previously, we have $I_{N+1} = m + k$, $I_{N+2} = m + l$). We can assume from the start that the weighting functions in the n-dimension $w_{n,i_n}(p_n)$, $i_n = 1, \ldots, I_n$, that is, the columns of $\mathbf{w}_n$ are linearly independent (in the sense of the $\mathfrak{L}_2[a_n, b_n]$ vector product) over the intervals $[a_n, b_n]$ for $n = 1, \ldots, N$. If not, we can always choose whatever number I_n of linearly independent functions from $w_{n,i_n}(p_n)$ and express the remaining as linear combinations of the independent ones.

Furthermore, we can assume that for all $n = 1, \ldots, N$, a set of orthonormal functions $\varphi_{n,i_n}(p_n)$, $i_n = 1, \ldots, I_n$ can be determined via the linear combination of the $w_{n,i_n}(p_n)$ by, for instance, the Gram–Schmidt type orthogonalization method. Thus, one can always convert (4.1) to the following equivalent form comprising an orthonormal system:

$$\begin{pmatrix} \dot{\mathbf{x}} \\ \mathbf{y} \end{pmatrix} = \left(C \underset{n=1}{\overset{N}{\boxtimes}} \varphi_n(p_n) \right) \begin{pmatrix} \mathbf{x} \\ \mathbf{u} \end{pmatrix}, \tag{4.2}$$

where $\varphi_{n,i_n}(p_n), n = 1, \ldots, I_n$, are the columns of $\varphi_n(p_n)$ forming a set of orthonormal functions over $[a_n, b_n]$ for all $n = 1, \ldots, N$. The tensor C is the corresponding result of the manipulations to S.

Corollary 4.1. *We can assume, without the loss of generality, that the functions* $w_{n,i_n}(p_n)$ *in the TP structure*

$$\mathbf{S}(\mathbf{p}) = S \underset{n=1}{\overset{N}{\boxtimes}} \mathbf{w}_n(p_n)$$

form an orthonormal system:

$$\forall n : \int_{a_n}^{b_n} w_{n,i}(p_n)w_{n,j}(p_n)dp_n = \delta_{i,j}, \quad 1 \le i, j \le I_n,$$

where $\delta_{i,j}$ *is the Kronecker delta function* ($\delta_{ij} = 1$, *if* $i = j$ *and* $\delta_{ij} = 0$, *if* $i \ne j$).

Now we execute compact higher-order singular value decomposition (CHOSVD) (see Definition 3.5) on the first N-dimension of the system tensor $S \in \mathbb{R}^{I_1 \times \cdots \times I_{N+2}}$ and obtain:

$$S = \mathcal{D} \underset{n=1}{\overset{N}{\boxtimes}} \mathbf{U}_n, \tag{4.3}$$

where $\mathbf{U}_n$ has the size of $I_n \times r_n$ and $r_n = rank_n(S)$. Further defining the following weighting functions as:

$$\tilde{\mathbf{w}}_n(p_n) = \mathbf{w}_n(p_n)\mathbf{U}_n$$

then leads to

$$\begin{pmatrix} \dot{\mathbf{x}} \\ \mathbf{y} \end{pmatrix} = \left(\mathcal{D} \underset{n=1}{\overset{N}{\boxtimes}} \tilde{\mathbf{w}}_n(p_n) \right) \begin{pmatrix} \mathbf{x} \\ \mathbf{u} \end{pmatrix}. \tag{4.4}$$

With $\mathbf{U}_n$ being orthonormal matrices and $w_{n,i_n}(p_n)$, $1 \le i_n \le I_n$ functions also forming an orthonormal set, the elements of the function $\tilde{\mathbf{w}}_n(p_n)$, namely, $\tilde{\mathbf{w}}_n(p_n) = (\tilde{w}_{n,1}(p_n), \cdots, \tilde{w}_{n,r_n}(p_n))$, also form an orthonormal system for all $n = 1, \ldots, N$.

Based on the above, we thus have the following theorem.

Theorem 4.1 (TP canonical form of qLPV models). *Assume a TP structure*

$$\begin{pmatrix} \dot{\mathbf{x}} \\ \mathbf{y} \end{pmatrix} = \left(S \underset{n=1}{\overset{N}{\boxtimes}} \mathbf{w}_n(p_n) \right) \begin{pmatrix} \mathbf{x} \\ \mathbf{u} \end{pmatrix}. \tag{4.5}$$

Based on Corollary 4.1 and via executing CHOSVD on S *(on its first N dimensions) one can determine*

$$\begin{pmatrix} \dot{\mathbf{x}} \\ \mathbf{y} \end{pmatrix} = \left(\mathcal{D} \underset{n=1}{\overset{N}{\boxtimes}} \tilde{\mathbf{w}}_n(p_n) \right) \begin{pmatrix} \mathbf{x} \\ \mathbf{u} \end{pmatrix}, \tag{4.6}$$

where tensor $\mathcal{D}$ *has a size of* $(r_1 \times \cdots \times r_N \times I_{N+1} \times I_{N+2})$, *with the following properties:*

1. The weighting functions $\tilde{w}_{n,i_n}(p_n)$, $i_n = 1, \ldots, r_n$ (termed as the i_nth singular function in the nth dimension, $n = 1, \ldots, N$) in vector $\tilde{\mathbf{w}}_n(p_n)$ form an orthonormal set:

$$\forall n : \int_{a_n}^{b_n} \tilde{w}_{n,i}(p_n)\tilde{w}_{n,j}(p_n)dp_n = \delta_{i,j}, \quad 1 \leq i, j \leq I_n,$$

where $\delta_{i,j}$ is the Kronecker delta function ($\delta_{ij} = 1$, if $i = j$ and $\delta_{ij} = 0$, if $i \neq j$).

2. The subtensors $\mathcal{D}_{i_n=i}$ have the properties of

    (a) all-orthogonality: two subtensors $\mathcal{D}_{i_n=i}$ and $\mathcal{D}_{i_n=j}$ are orthogonal for all possible values of n, i, and j : $\langle \mathcal{D}_{i_n=i}, \mathcal{D}_{i_n=j} \rangle = 0$ when $i \neq j$,

    (b) ordering: $\|\mathcal{D}_{i_n=1}\| \geq \|\mathcal{D}_{i_n=2}\| \geq \ldots \geq \|\mathcal{D}_{i_n=r_n}\| > 0$ for all possible values of $n = 1, \ldots, N + 2$.

3. The Frobenius-norm $\|\mathcal{D}_{i_n=i}\|$, symbolized by $\sigma_i^{(n)}$, are n-mode singular values of $\mathcal{D}$.

4. $\mathcal{D}$ is termed core tensor consisting of the linear time-invariant (LTI) vertex systems.

The form (4.4), or equivalently, (4.6), constitutes the TP canonical form of the qLPV model. Note that the TP canonical model has a TP structure but is no longer a TP model as originally started in (4.1). The reason is that the canonical model features orthonormal weighting functions, which means that it must not be able to satisfy the polytopic conditions and hence is not polytopic (see Definition 2.4 for TP model). Chapter 6 presents the incorporation of convexity conditions to convert a canonical model to become a TP model.

Remark 4.1. If there are equal singular values on any dimensions when CHOSVD is executed, then the canonical form is not unique. In this case the n-mode singular vectors corresponding to the same n-mode singular value can be replaced by any of their orthonormal linear combinations (see [DLDMV00]).

4.2 Numerical reconstruction

This section is based on the works of [BSV06], [SV09]. It shows that the TP model transformation in Chapter 3, namely, the three steps of discretization, higher-order singular value decomposition (HOSVD), and structure extraction, is able to numerically reconstruct the TP canonical form of (4.6). It will be proven that if the discretization density is increased to infinity, the outcome of the TP model transformation process (3.18) will converge exactly to (4.6). The content of this section is highly mathematical. Readers who are less interested in the mathematical details may skip it and move to the next section, where an example of the numerical reconstruction is presented.

Let us divide the intervals $[a_n, b_n]$, $n = 1, \ldots, N$ into M_n number of disjunct subintervals Ξ_{n,m_n}, $1 \leq m_n \leq M_n$ so that:

$$a_n = \xi_{n,0} < \xi_{n,1} < \ldots < \xi_{n,M_n} = b_n,$$

$$
\begin{aligned}
\Xi_{n,m_n} &= [\xi_{n,m_n-1}, \xi_{n,m_n}), 1 \leq m_n \leq M_N - 1, \\
\Xi_{n,M_N} &= [\xi_{n,M_N-1}, \xi_{n,M_N}].
\end{aligned}
$$

Utilizing the above intervals, we can discretize the function $\mathbf{S}(\mathbf{p})$ at given points over the intervals. Let

$$x_{n,m_n} \in \Xi_{n,m_n}, \quad 1 \leq m_n \leq M_n, 1 \leq n \leq N. \tag{4.7}$$

For brevity, we denote $M_{N+1} = I_{N+1}$, $M_{N+2} = I_{N+2}$. Let us introduce the discrepancy functions described by the sequences x_{n,m_n} as follows:

$$\Delta_n(s) = \rho_n \left(\sum_{k=1}^{M_n} I(x_{n,k} < s) \right) - (s - a_n),$$

$a_n \leq s < b_n$, $\Delta_n(b_n) = 0$, where I is the indicator function and

$$\rho_n = \frac{b_n - a_n}{M_n},$$

and denote

$$\Delta_n = \sup_{a_n \leq s < b_n} |\Delta_n(s)|, \quad \Delta_{2,n} = \left(\int_{a_n}^{b_n} \Delta_n^2(s) ds \right)^{1/2}.$$

The elements x_{n,m_n} of (4.7) define a hyper-rectangular grid $\mathbf{g}$, whose elements are $\mathbf{g}_{m_1,\ldots,m_N} = \left(x_{1,m_1} \quad \cdots \quad x_{N,m_N} \right)$. Let us discretize the matrix function $\mathbf{S}(\mathbf{p})$ for the grid points in $\mathbf{g}$:

$$\mathbf{B}_{m_1,m_2,\ldots,m_N} = \mathbf{S}(\mathbf{g}_{m_1,m_2,\ldots,m_N}), \quad 1 \leq m_n \leq M_n, 1 \leq n \leq N$$

and construct an $(N + 2)$-dimensional tensor $\mathcal{B}$ with the matrices $\mathbf{B}_{m_1,\ldots,m_N}$ thus obtained. Obviously, the size of this tensor is $M_1 \times \cdots \times M_{N+2}$. Further, we discretize the vector valued functions $\mathbf{w}_n(p_n)$ over the discretization points x_{n,m_n} and construct matrices $\mathbf{W}^{(n)} \in \mathbb{R}^{M_n \times r_n}$ from the discretized values as:

$$\mathbf{W}^{(n)} = \begin{pmatrix} w_{n,1}(x_{n,1}) & w_{n,2}(x_{n,1}) & \cdots & w_{n,r_n}(x_{n,1}) \\ w_{n,1}(x_{n,2}) & w_{n,2}(x_{n,2}) & \cdots & w_{n,r_n}(x_{n,2}) \\ \vdots & & \ddots & \vdots \\ w_{n,1}(x_{n,M_n}) & w_{n,2}(x_{n,M_n}) & \cdots & w_{n,r_n}(x_{n,M_n}) \end{pmatrix}. \tag{4.8}$$

Let us denote $w_{i,k}^{(n)} = w_{n,i}(x_{n,k}), 1 \leq k \leq M_n, 1 \leq i \leq r_n, 1 \leq n \leq N$, then we can write $\mathbf{W}^{(n)} = \left(\mathbf{W}_1^{(n)} \quad \mathbf{W}_2^{(n)} \quad \cdots \quad \mathbf{W}_{r_n}^{(n)} \right) \in \mathbb{R}^{M_n \times r_n}, 1 \leq n \leq N$, where $\mathbf{W}_i^{(n)} = \left(w_{i,1}^{(n)}, \ldots, w_{i,M_n}^{(n)} \right)^T, 1 \leq i \leq r_n$ denote the column vectors of the matrix $\mathbf{W}^{(n)}$. Then tensor $\mathcal{B}$ can simply be given by (4.3) and (4.4) as

$$\mathcal{B} = \mathcal{D}_0 \overset{N}{\underset{n=1}{\boxtimes}} \mathbf{W}^{(n)}. \tag{4.9}$$

Let us denote the matrices $\varepsilon_n \in \mathbb{R}^{r_n \times r_n}, 1 \leq n \leq N$,

$$\varepsilon_n = \left(\varepsilon_{i,j}^{(n)} \right)_{i,j=1}^{r_n},$$

where

$$\varepsilon_{i,j}^{(n)} = \delta_{i,j} - \rho_n \sum_{k=1}^{M_n} w_{n,i}(x_{n,k}) w_{n,j}(x_{n,k}), \quad 1 \leq i, j \leq r_n.$$

The following lemma gives an estimation for the upper bound for the quantity $\|\varepsilon_n\|$, which guarantees the convergence $\|\varepsilon_n\| \to 0$ if $\Delta_n \to 0$ and which plays a basic role in the formulation of our results.

Let us also assume that the functions $w_{n,i_n}(p_n), 1 \leq i_n \leq r_n \ (1 \leq n \leq N)$ are piecewise continuously differentiable on the interval $p_n \in [a_n, b_n]$ (at the end points of the interval we understood left and right hand side derivatives). Let us denote

$$K_{0,n} = \max_{1 \leq i \leq r_n} \max_{a_n \leq s \leq b_n} \left| w_{n,i}(s) \right|$$

$$K_{1,n} = \max_{1 \leq i \leq r_n} \int_{a_n}^{b_n} \left| (w_{n,i}(s))' \right| ds,$$

$$K_{2,n} = \max_{1 \leq i \leq r_n} \left(\int_{a_n}^{b_n} \left[(w_{n,i}(s))' \right]^2 ds \right)^{1/2}.$$

Lemma 1. The Frobenius-norm of the matrix $\varepsilon_n = \mathbf{E}_{r_n} - \frac{b_n - a_n}{M_n} \mathbf{W}^{(n)T} \mathbf{W}^{(n)}$ satisfies the inequalities

$$\|\varepsilon_n\| \leq 2 r_n K_{0,n} K_{1,n} \Delta_n$$

and

$$\|\varepsilon_n\| \leq r_n K_{2,n} \Delta_{2,n}.$$

Corollary 4.2. *If the intervals $[a_n, b_n]$ are divided into equidistant subintervals, that is, in the case of*

$$\xi_{n,k} = a_n + k \frac{b_n - a_n}{M_n}, \quad 1 \leq k \leq M_n,$$

then the following inequalities hold:

$$\|\varepsilon_n\| \leq 2r_n K_{0,n} K_{1,n} \frac{b_n - a_n}{M_n}$$

and

$$\|\varepsilon_n\| \leq 2r_n K_{0,n} K_{2,n} \frac{(b_n - a_n)^{3/2}}{M_n}.$$

The statements of Lemma 1 follow from the results of Lemma 2 (the Koksma–Hlawka inequality and its $\mathfrak{L}_2$ version in the one-dimension case).

Lemma 2. Let $y_1, \ldots, y_L$ be arbitrary points from the interval $[a, b]$. If the function $u(t)$ ($t \in [a, b]$) is piecewise continuously differentiable (at the end points of the interval we have left and right hand side derivatives), then the following inequalities hold:

$$\left| \frac{b - a}{L} \sum_{l=1}^{L} u(y_l) - \int_a^b u(s)ds \right| \leq \sup_{a \leq s \leq b} |\Delta_L(s)| \int_a^b |u'(s)| \, ds,$$

$$\left| \frac{b - a}{L} \sum_{l=1}^{L} u(y_l) - \int_a^b u(s)ds \right| \leq \left(\int_a^b (\Delta_L(s))^2 \, ds \right)^{1/2} \left(\int_a^b (u'(s))^2 \, ds \right)^{1/2},$$

where $\Delta_L(s) = \frac{b-a}{L} \left(\sum_{l=1}^{L} I(y_l < s) \right) - (s - a), a \leq s \leq b.$

Proof. Since $u(y) = u(b) - \int_a^b u'(s)I(y < s)ds$ and with integrating by part we get

$$\int_a^b u(s)ds = u(b)b - u(a)a - \int_a^b u'(s)sds,$$

then

$$\frac{b - a}{L} \sum_{l=1}^{L} u(y_l) = (b - a)u(b) - \frac{b - a}{L} \int_a^b u'(s) \sum_{l=1}^{L} I(y_l < s)ds$$

$$= -\int_a^b u'(s)\Delta_L(s)ds + \int_a^b u(s)ds$$

and

$$\left| \frac{b-a}{L} \sum_{i=1}^{L} u(y_l) - \int_a^b u(s)ds \right| \leq \left| \int_a^b u'(s)\Delta_L(s)ds \right|.$$

From this relation the first inequality of Lemma 2 immediately follows, and we get the second inequality by use of the Cauchy–Schwartz inequality. □

Now for the proof of Lemma 1, since the functions $w_{n,i}(x_n), 1 \leq i \leq r_n$ are orthonormal in the $\mathcal{L}_2$ sense on the interval $[a_n, b_n]$, therefore

$$\int_{a_n}^{b_n} w_{n,i}(s)w_{n,j}(s)ds = \delta_{i,j}.$$

By the first inequality of Lemma 2, it follows that

$$|\varepsilon_{ij}| \leq \Delta_n \left| \int_a^b \left(w_{n,i}(s)w_{n,j}(s) \right)' ds \right|$$

$$\leq \Delta_n \int_a^b \left| (w_{n,i}(s))' \, w_{n,j}(s) + w_{n,i}(s) \left(w_{n,j}(s) \right)' \right| ds$$

$$\leq 2\Delta_n K_{0,n} K_{1,n}.$$

We can easily get the second inequality of Lemma 1

$$\|\varepsilon_n\| = \left(\sum_{i,j=1}^{r_n} \varepsilon_{i,j}^2 \right)^{1/2} \leq \left(r_n^2 4\Delta_n^2 K_{0,n}^2 K_{1,n}^2 \right)^{1/2}$$

$$= 2r_n \Delta_n K_{0,n} K_{1,n}.$$

For simplicity, we assume in the following that the singular values are all different for each dimension of the discretized tensor.

Based on the previous notations, the discretized tensor can be written in the form $\mathcal{B} = \mathbb{R}^{M_1 \times \cdots \times M_{N+2}}$, where

$$b_{m_1,\dots,m_{N+2}} = \sum_{i_1=1}^{r_1} \cdots \sum_{i_N=1}^{r_N} d_{i_1,\dots,i_N,m_{N+1},m_{N+2}} w_{i_1,m_1}^{(1)} \cdots w_{i_N,m_N}^{(N)}, \tag{4.10}$$

for $1 \leq m_n \leq M_n, 1 \leq n \leq N+2$. Let us consider the discretized tensor

$$\mathcal{B} = \mathcal{D}_0 \overset{N}{\underset{n=1}{\boxtimes}} \mathbf{W}^{(n)} \in \mathbb{R}^{M_1 \times \cdots \times M_{N+2}}. \tag{4.11}$$

By the HOSVD decomposition of the discretized tensor we arrive at:

$$\mathcal{B} = \mathcal{D}^d \overset{N+2}{\underset{n=1}{\boxtimes}} \mathbf{U}^{(n)} \tag{4.12}$$

where $\mathcal{D}^d$ is the so-called core tensor, and $\mathbf{U}^{(n)} = \begin{pmatrix} \mathbf{U}_1^{(n)} & \mathbf{U}_2^{(n)} & \cdots & \mathbf{U}_{M_n}^{(n)} \end{pmatrix}$ is an $M_n \times M_n$-size orthogonal matrix ($1 \leq n \leq N + 2$).

The subtensors $\mathcal{D}_{\alpha}^{d,n}$ of the tensor $\mathcal{D}^d \in \mathbb{R}^{M_1 \times \cdots \times M_{N+2}}$ (where the nth index $\alpha = 1, \ldots, M_n$ is fixed) has the following properties:

1. Tensors $\mathcal{D}_{\alpha}^{d,n}$ and $\mathcal{D}_{\beta}^{d,n}$ are orthogonal for all $1 \leq n \leq N + 2$ and $\alpha \neq \beta$, namely

$$\langle \mathcal{D}_{\alpha}^{d,n}, \mathcal{D}_{\beta}^{d,n} \rangle = 0 \tag{4.13}$$

2. $\sigma_1^{d,n} \geq \cdots \geq \sigma_{M_n}^{d,n}$, where $\sigma_i^{d,n} = \left\| \mathcal{D}_i^{d,n} \right\|$, $1 \leq i \leq M_n$, $1 \leq n \leq N + 2$.

3. We need the n-mode matrix unfolding of discretized tensor $\mathcal{B}$,

$$\mathbf{B}_{(n)}^d = \mathbf{U}^{(n)} \mathbf{D}_{(n)}^d \left(\mathbf{U}^{(n+1)} \otimes \cdots \mathbf{U}^{(N+2)} \otimes \cdots \otimes \mathbf{U}^{(1)} \otimes \cdots \otimes \mathbf{U}^{(n-1)} \right)^T, \tag{4.14}$$

where $\otimes$ denotes the Kronecker product. Let r_n^d be the rank of matrix $\mathbf{B}_{(n)}^d$. Then $r_n^d = \mathrm{rank}(\mathbf{B}_{(n)}^d) = \mathrm{rank}_n(\mathcal{B})$ is the dimension of the linear space spanned by the n-mode vectors. The orthogonal matrix $\mathbf{D}_{(n)}^d$ is the singular matrix from the singular value decomposition (SVD) decomposition with the positive $\sigma_1^{d,n} \geq \cdots \geq \sigma_{r_n^d}^{d,n} > 0$ singular values in its diagonal. The other singular values and the rest of the elements of $\mathbf{D}_{(n)}^d$ are 0. The Frobenius-norm $\sigma_i^{d,n} = \left\| \mathcal{D}_i^{d,n} \right\|$ is equal to the n-mode singular values of tensor $\mathcal{B}$, and vectors $\mathbf{U}_i^{(n)}$, $1 \leq i \leq M_n$ are n-mode singular vectors. Then it is true for the n-mode singular values

$$\|\mathcal{B}\|^2 = \sum_{i=1}^{r_1} \left(\sigma_i^{d,1} \right)^2 = \cdots = \sum_{i=1}^{r_{N+2}} \left(\sigma_i^{d,N+2} \right)^2. \tag{4.15}$$

4. The core tensor is

$$\mathcal{D}^d = \mathcal{B} \overset{N+2}{\underset{n=1}{\boxtimes}} \mathbf{U}^{(n)T}. \tag{4.16}$$

Let $R_n = M_{n+1} \ldots M_{N+2} M_1 \ldots N_{n-1}$. By definition for the $(i_n, k_{n,m})$th element of matrix $\mathbf{B}_{(n)}^d \in \mathbb{R}^{M_n \times R_n}$, the relation $b_{i_n, k_{n,m}}^{(n)} = b_{m_1, \ldots, m_N}$ is true, where

$$k_{n,m} = (m_{n+1} - 1)M_{n+2} \ldots M_{N+2} M_1 \ldots M_{n-1} + \ldots$$
$$+ (m_{N+2} - 1)M_1 \ldots M_{n-1} + (m_1 - 1)M_2 \ldots M_{n-1} + \ldots$$
$$+ (m_{n-2} - 1)M_{n-1} + m_{n-1}$$
$$m = (m_1, \ldots, m_{N+2}), 1 \leq m_n \leq M_n, 1 \leq n \leq N + 2.$$

Note that the value of $k_{n,m}$ only depends on the values of $m_1, \ldots, m_{n-1}, m_{n+1}, \ldots, m_N$, and not on the actual value of index i_n.

Lemma 3. For every matrix $\mathbf{U} \in \mathbb{R}^{M_n \times L_n}$ with $rank(\mathbf{U}) = L_n$, and for every tensor $\mathcal{A} \in \mathbb{R}^{L_1 \times \cdots \times L_N}$, the following relations hold:

$$\operatorname*{rank}_k(\mathcal{A} \times_n \mathbf{U}) = \operatorname*{rank}_k(\mathcal{A}), \quad 1 \le k \le N.$$

Proof. Let us denote $r_k = \operatorname{rank}_k \mathcal{A}$ and

$$\mathbf{A}_{(k)}(i_1, \ldots, i_{k-1}, i_{k+1}, \ldots, i_N) = \begin{pmatrix} a_{i_1, \ldots, i_{k-1}, 1, i_{k+1}, \ldots, i_N} \\ \vdots \\ a_{i_1, \ldots, i_{k-1}, L_k, i_{k+1}, \ldots, i_N} \end{pmatrix},$$

$k = 1 \ldots N$. First, we prove that $\operatorname{rank}_n(\mathcal{A} \times_n \mathbf{U}) = r_n$. By the definition of the n-mode product, we can write

$$(\mathcal{A} \times_n \mathbf{U})_{i_1, \ldots, i_{n-1}, j_n, i_{n+1}, \ldots, i_N} = \sum_{i_n=1}^{L_n} a_{i_1, \ldots, i_N} u_{j_n, i_n},$$

therefore the M_n-dimensional n-mode vectors of the tensor $(\mathcal{A} \times_n \mathbf{U})$ can be given in the form

$$\sum_{i_n=1}^{L_n} a_{i_1, \ldots, i_n, \ldots, i_N} u_{i_n} = \mathbf{U}\mathbf{A}_{(n)}(i_1, \ldots, i_{n-1}, i_{n+1}, \ldots, i_N),$$

where $u_i, 1 \le i \le L_n$ denotes the column vectors of the matrix $\mathbf{U}$. Since $r_n = \operatorname{rank}_n \mathcal{A}$ and the vectors $\mathbf{A}_{(n)}(i_1, \ldots, i_{n-1}, i_{n+1}, \ldots, i_N), 1 \le i_j \le L_j, j \ne n$ from the right side of the last equation are the n-mode vectors of the tensor $\mathcal{A}$, then we can select from them exactly r_n linearly independent column vectors. From the condition $\operatorname{rank} \mathbf{U} = L_n \ge r_n$ it follows that the rank of linear space spanned by the n-mode vectors is exactly r_n.

Proof of the case of $\operatorname{rank}_k(\mathcal{A} \times_n \mathbf{U}) = \operatorname{rank}_k \mathcal{A}, 1 \le k \le N, k \ne n$. Let us denote the $L_k \times L_n$ matrix

$$\mathbf{A}_{(k,n)} = \left[\mathbf{A}_{(k)}(i_1, \ldots, i_{k-1}, i_{k+1}, \ldots, i_{n-1}, j, i_{n+1}, \ldots, i_N) \right],$$

$j = 1, \ldots, L_n$. It can be seen that the k-mode vectors of the tensor $(\mathcal{A} \times_n \mathbf{U})$ correspond to the column vectors of the matrices $\mathbf{A}_{(k,n)} \mathbf{U}^T, n = 1, \ldots, L_N$. Since $\operatorname{rank}_k \mathcal{A} = r_k$, the set of vectors $\mathbf{A}_{(k)}(i_1, \ldots, i_{n-1}, i_{n+1}, \ldots, i_N), 1 \le i_j \le L_j, j \ne n$ consists of r_k linearly independent vectors. Thus, from the condition $\operatorname{rank} \mathbf{U} = L_n \ge r_n$, it follows that the rank of linear space generated by the column vectors of the matrices $\mathbf{A}_{(k,n)} \mathbf{U}^T, n = 1, \ldots, L_N$ is exactly r_k too. $\square$

We note that the results of Lemma 3 can be derived from Theorem 6 of [MS74].

By the definition of the matrices ε_n, $1 \leq n \leq N$ we have $\mathbf{W}^{(n)^T} \mathbf{W}^{(n)} = E_{r_n} - \varepsilon_n$. If $\|\varepsilon_n\| < 1$, $1 \leq n \leq N$, then by the well-known result of matrix theory it follows that

$$\text{rank}(E_{r_n} - \varepsilon_n) = \text{rank}(\mathbf{W}^{(n)T} \mathbf{W}^{(n)}) = r_n, 1 \leq n \leq N,$$

thus using Lemma 3, we get $r_k^d = \text{rank}_k \mathcal{B} = r_k$. As a consequence of this relation, we have the following theorem.

Theorem 4.2. *If* $\|\varepsilon_n\| < 1$, $1 \leq n \leq N$, *then* $r_k^d = \text{rank}_k \mathcal{B} = \text{rank}_k \mathcal{D}^d = r_k$, $1 \leq n \leq N$.

Note that during the proof of the following theorem, it is found that $r_n^d = r_n$, $1 \leq n \leq N$, if $\Delta = \min_{1 \leq n \leq N} \Delta_n$ is small enough.

Consider the $r_1 \times \cdots \times r_{N+2}$-size reduced version $\widetilde{\mathcal{D}}^d = (\mathcal{D}_{m_1 \ldots m_N+2}^d, 1 \leq m_n \leq r_n, 1 \leq n \leq N + 2)$ of the $M_1 \times \cdots \times M_{N+2}$-size tensor $\mathcal{D}^d$.

It follows from (4.2) that $\sigma_{r_n+1}^{d,n} = \cdots = \sigma_{M_n}^{d,n}$; thus, instead of $\mathcal{D}^d$ the analysis of tensor $\widetilde{\mathcal{D}}^d$ is enough.

Theorem 4.3. *If* $\Delta \to 0$ *then* $\sqrt{\rho} \widetilde{\mathcal{D}}^d \to \mathcal{D}$ *and* $\mathbf{U}^{(n)} \to \mathbf{U}_n$, $n = N + 1, N + 2$, *where*

$$\rho = \prod_{n=1}^{N} \rho_n = \prod_{n=1}^{N} \frac{b_n - a_n}{M_n}.$$

Proof. Taking into consideration the result of the discretization in two different ways (4.11) and (4.12), we get

$$\mathcal{B} = \mathcal{D}_0 \underset{n=1}{\overset{N}{\boxtimes}} \mathbf{W}^{(n)} = \mathcal{D}^d \underset{n=1}{\overset{N+2}{\boxtimes}} \mathbf{U}^{(n)}. \tag{4.17}$$

Then, by applying the rule of n-mode multiplication of tensor by matrices:

$$\mathcal{D}_0 \underset{n=1}{\overset{N}{\boxtimes}} \mathbf{U}^{(n)^T} \mathbf{W}^{(n)} = \mathcal{D}^d \underset{n=1}{\overset{N+2}{\boxtimes}} \mathbf{U}^{(n)^T} \mathbf{U}^{(n)}. \tag{4.18}$$

Here $\mathbf{U}^{(n)}$ is an $M_n \times M_n$ orthogonal matrix; then

$$\mathbf{U}^{(n)^T} \mathbf{U}^{(n)} = E_{M_n},$$

so from (4.17) it follows that

$$\mathcal{D} \times_{N+1} \mathbf{H}^{(N+1)} \times_{N+2} \mathbf{H}^{(N+2)} \underset{n=1}{\overset{N}{\boxtimes}} \mathbf{H}^{(n)} = \sqrt{\rho} \mathcal{D}^d, \tag{4.19}$$

where $\mathbf{H}^{(n)} = \sqrt{\rho_n} \mathbf{U}^{(n)^T} \mathbf{W}^{(n)} \in \mathbb{R}^{M_n \times r_n}$, $1 \leq n \leq N$ and $\mathbf{H}^{(n)} = \mathbf{U}^{(n)^T} \mathbf{U}_n \in \mathbb{R}^{I_n \times I_n}$, $n = N + 1, N + 2$. It is evident that

$$\mathbf{H}^{(n)^T} \mathbf{H}^{(n)} = \rho_n \mathbf{W}^{(n)^T} \mathbf{W}^{(n)} = E_{r_n} - \varepsilon_n, 1 \leq n \leq N, \tag{4.20}$$

where $\|\varepsilon_n\| \to 0$, if $\Delta \to 0$; thus $\text{rank}(\mathbf{H}^{(n)}) = r_n$ if Δ is small enough. We note that the matrices $\mathbf{H}^{(N+1)}$ and $\mathbf{H}^{(N+2)}$ are orthogonal, that is,

$$\mathbf{H}^{(n)^T} \mathbf{H}^n = \mathbf{E}_{I_n}, n = N+1, N+2, \tag{4.21}$$

which immediately follows from the property of $\mathbf{U}^{(n)}$ and $\mathbf{W}_n$.

Let us denote the column vectors of matrix $\mathbf{H}^{(n)}$ by $\mathbf{H}_j^{(n)}, 1 \leq j \leq r_n$. The reduced $r_n \times r_n$ matrix derived from $\mathbf{H}^{(n)}, 1 \leq n \leq N$ is denoted by $\widetilde{\mathbf{H}}^{(n)}$ (we select the first r_n row of matrix $\mathbf{H}^{(n)}$). From (4.17) it follows that the linear spaces spanned by the linearly independent vectors $\mathbf{W}_1^{(n)}, ..., \mathbf{W}_{r_n}^{(n)}$ (if Δ is small enough) and $\mathbf{U}_1^{(n)}, ..., \mathbf{U}_{r_n}^{(n)}$, respectively, are the same; therefore, vectors $\mathbf{W}_i^{(n)}$ and $\mathbf{U}_j^{(n)}$ are orthogonal, if $1 \leq i \leq r_n, r_n + 1 \leq j \leq M_n$; then

$$\mathbf{H}^{(n)^T} \mathbf{H}^{(n)} = \widetilde{\mathbf{H}}^{(n)^T} \widetilde{\mathbf{H}}^{(n)}. \tag{4.22}$$

According to (4.20) and (4.22), the column vectors $\widetilde{\mathbf{H}}_j^{(n)}, 1 \leq j \leq r_n$ are asymptotically orthonormal, namely $\widetilde{\mathbf{H}}_j^{(n)^T} \widetilde{\mathbf{H}}_j^{(n)} \to \delta_{ij}, \Delta \to 0, 1 \leq i, j \leq r_n$; thus

$$\widetilde{\mathbf{H}}^{(n)^T} \widetilde{\mathbf{H}}^{(n)} \to \mathbf{E}_{I_n}, \quad \Delta \to 0. \tag{4.23}$$

Let us increase the discretization points $a_n \leq x_{n,1} < \cdots < x_{n,M_n} \leq b_n, n = 1, ..., N$ with the property $\Delta \to 0$. Then the elements of matrix $\widetilde{\mathbf{H}}^{(n)}$ are still bounded, so a partial series can be selected and a matrix $\mathbf{G_n}, \mathbf{G}_n^T \mathbf{G} = \mathbf{E}_{r_n}$ can be given independently from the discretization points in such a way that the convergence $\widetilde{\mathbf{H}}^{(n)} \to \mathbf{G}_n$ is valid for the partial series. So, based on the partial series and an appropriate tensor $\widetilde{\mathcal{D}}$,

$$\sqrt{\rho\widetilde{\mathcal{D}}^d} \to \mathcal{D} \times_1 \mathbf{G}_1 \cdots \times_{N+2} \mathbf{G}_{N+2} = \widetilde{\mathcal{D}}.$$

Here $\widetilde{\mathcal{D}}$ satisfies the properties of a core tensor because it is given as the limit value of core tensors. By rewriting this equation we get

$$\mathcal{D} \overset{N+2}{\underset{n=1}{\boxtimes}} \mathbf{G}_n = \widetilde{\mathcal{D}} \tag{4.24}$$

As the construction of $\mathcal{D}$ by (4.1) is unique, then the matrices $\mathbf{G}_n, 1 \leq n \leq N$ and tensor $\widetilde{\mathcal{D}}$ are also unique. Then it is evident that $\mathbf{G}_n = \mathbf{E}_{r_n}, \widetilde{\mathcal{D}} = \mathcal{D}$, and

$$\widetilde{\mathbf{H}}^{(n)} \to \mathbf{E}_{r_n}, \quad \sqrt{\rho\widetilde{\mathcal{D}}^d} \to \widetilde{\mathcal{D}}, \quad \Delta \to 0 \tag{4.25}$$

and

$$\mathbf{U}^{(n)} \to \mathbf{U}_n, n = N+1, N+2 \tag{4.26}$$

$\square$

Let us denote the elements of matrix $\mathbf{U}^{(n)}$ by $\mathbf{U}^{(n)}_{i,k}$ and introduce the step functions $u_{n,i}(x)$, $1 \le i \le r_n$, which are similar to $v_{n,i}(x)$ in such a way that the interval $\Xi_{n,k}$ is

$$u_{n,i}(x) = \frac{1}{\sqrt{\rho}} \mathbf{U}^{(n)}_{i,k} I(x \in \Xi_{n,k}), \quad 1 \le k \le M_n \tag{4.27}$$

$$(v_{n,i}(x) = w^{(n)}_{i,k} I(x \in \Xi_{n,k})).$$

Theorem 4.4. *If* $\Delta \to 0$, *then*

$$\int_{a_n}^{b_n} (w_{n,i}(x) - u_{n,i}(x))^2 dx \to 0, \quad 1 \le i \le r_n, 1 \le n \le N \tag{4.28}$$

Proof. It is trivial that

$$\int_{a_n}^{b_n} (w_{n,i}(x) - u_{n,i}(x))^2 dx \le$$

$$\le 2 \int_{a_n}^{b_n} (w_{n,i}(x) - v_{n,i}(x))^2 dx + 2 \int_{a_n}^{b_n} (v_{n,i}(x) - u_{n,i}(x))^2 dx,$$

so it is enough to analyze the second integral on the right. Then from Lemma 1 we get

$$\int_{a_n}^{b_n} (v_{n,i}(x) - u_{n,i}(x))^2 dx = \sum_{k=1}^{M_n} \left(w^{(n)}_{n,i} - \frac{1}{\sqrt{\rho}} U^{(n)}_{i,k} \right)^2 \rho_n =$$

$$= \sum_{k=1}^{M_n} \left[(w^{(n)}_{i,k})^2 \rho_n - 2\sqrt{\rho_n} w^{(n)}_{i,k} U^{(n)}_{i,k} + (U^{(n)}_{i,k})^2 \right] \to 0, \Delta \to 0.$$

$\square$

4.3   The TORA example

The TP canonical form may be derived analytically in some special cases. As the model embeds greater complexity, however, the analytical derivation of going through the orthonormalization of weighting functions and HOSVD on $\mathcal{S}$ can easily become a highly tedious, if not impossible, task. On the other hand, the TP model transformation of Chapter 3 can be executable on a regular PC in a few minutes regardless of the actual dynamic complexity of the given qLPV model. Execution is also independent of the type of model available, be it of state space equation, fuzzy rule-based, neural network, or even table look-up. The purpose of this section is to illustrate the application of the TP model transformation to obtain a TP canonical model.

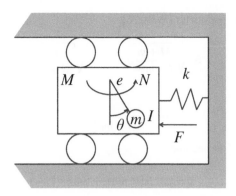

Figure 4.1: TORA system. (Modified from [BSVY06]–P. Baranyi, L. Szeidl, and P. Várlaki, "Numerical reconstruction of the HOSVD based canonical form of poly-topic dynamic models," Proceedings of the *10th International Conference on Intelligent Engineering Systems*, pages 196-201, London, UK, June 26-28, 2006.)

The example we choose is the translational oscillations with an eccentric rotational proof mass actuator (TORA) system, which was originally proposed as a simplified model of a dual-spin spacecraft with mass imbalance for investigating the resonance capture phenomenon [RKM92], [JFK96]. The same system was later studied relating to the use of rotational proof-mass actuator for feedback stabilization of translational motion [WBC96], [BBC94]. The system was also investigated as a fourth-order benchmark problem for nonlinear control design [WB92], [BBC95], [BBC98].

4.3.1 Equations of motion

The TORA system is shown in Figure 4.1. A cart of mass M is connected to a fixed wall by a linear spring of stiffness k. The cart is constrained to have one-dimensional travel q in the horizontal plane. F is the disturbance force on the cart. Attached to the cart is a rotating proof-mass actuator with moment of inertia I and a distance e between the rotation point and the center of the proof mass m. The control torque N is applied to the proof mass. θ is the angular position of the rotational proof mass: $\theta = 0°$ is perpendicular to the motion of the cart, while $\theta = 90°$ is aligned with the positive q direction. The equations of motion are given by [Ber98]:

$$(M + m)\ddot{q} + kq = -me(\ddot{\theta}\cos\theta - \dot{\theta}^2\sin\theta), \tag{4.29}$$

$$(I + me^2)\ddot{\theta} = -me\ddot{q}\cos\theta + N. \tag{4.30}$$

With the normalized parameter [WBC96]:

$$\xi \simeq \sqrt{\frac{M+m}{I+me^2}}q, \quad \tau \simeq \sqrt{\frac{k}{M+m}}t, \tag{4.31}$$

and

$$u \simeq \frac{M+m}{k(I+me^2)}N, \tag{4.32}$$

the equations of motion become

$$\ddot{\xi} + \xi = \rho\left(\dot{\theta}^2\sin\theta - \ddot{\theta}\cos\theta\right), \tag{4.33}$$

$$\ddot{\theta} = -\rho\ddot{\xi}\cos\theta + u, \tag{4.34}$$

where ξ is the normalized cart position, u is the per unit control torque, τ is the normalized time, and differentiation is based on τ. The quantity ρ is the coupling between the rotational and the translational motions:

$$\rho \simeq \frac{me}{\sqrt{(I+me^2)(M+m)}}. \tag{4.35}$$

The above equations can be put in the state space model form

$$\dot{\mathbf{x}}(t) = f(\mathbf{x}(t)) + g(\mathbf{x}(t))u(t), \tag{4.36}$$

$$y(t) = c(\mathbf{x}(t)),$$

where

$$f(\mathbf{x}(t)) = \begin{pmatrix} x_2(t) \\ \frac{-x_1(t)+\rho x_4^2(t)\sin x_3(t)}{1-\rho^2\cos^2 x_3(t)} \\ x_4(t) \\ \frac{\rho\cos x_3(t)(x_1(t)-\rho x_4^2(t)\sin x_3(t))}{1-\rho^2\cos^2 x_3(t)} \end{pmatrix}, \quad g(\mathbf{x}(t)) = \begin{pmatrix} 0 \\ \frac{-\rho\cos x_3(t)}{1-\rho^2\cos^2 x_3(t)} \\ 0 \\ \frac{1}{1-\rho^2\cos^2 x_3(t)} \end{pmatrix}, \tag{4.37}$$

$$c(\mathbf{x}(t)) = \begin{pmatrix} 0 & 0 & x_3(t) & 0 \\ 0 & 0 & 0 & x_4(t) \end{pmatrix},$$

with $\mathbf{x}(t) = \begin{pmatrix} x_1(t) & x_2(t) & x_3(t) & x_4(t) \end{pmatrix}^T = \begin{pmatrix} \xi & \dot{\xi} & \theta & \dot{\theta} \end{pmatrix}^T$. Let us write the above equation in the typical form of the qLPV state space model as

$$\dot{\mathbf{x}}(t) = \mathbf{S}(\mathbf{p}(t))\begin{pmatrix} \mathbf{x}(t) \\ u(t) \end{pmatrix} \quad y = \mathbf{C}\mathbf{x}(t), \tag{4.38}$$

where system matrix $\mathbf{S}(\mathbf{p}(t))$ contains:

$$\mathbf{S}(\mathbf{p}(t)) = \begin{pmatrix} \mathbf{A}(\mathbf{p}(t)) & \mathbf{B}(\mathbf{p}(t)) \end{pmatrix}$$

Table 4.1: Parameters of the TORA System

Description	Parameter	Value	Units
Cart mass	M	1.3608	kg
Arm mass	m	0.096	kg
Arm eccentricity	e	0.0592	m
Arm inertia	I	0.0002175	kg m^2
Spring stiffness	k	186.3	N/m
Coupling parameter	ρ	0.200	—

and $\mathbf{p}(t) = \big(x_3(t) \quad x_4(t)\big) \in \Omega$ is a time-varying parameter vector, thus

$$\mathbf{A}(x_3(t), x_4(t)) = \begin{pmatrix} 0 & 1 & 0 & 0 \\ -\frac{1}{1-\rho^2\cos^2 x_3(t)} & 0 & 0 & \frac{\rho x_4(t)\sin x_3(t)}{1-\rho^2\cos^2 x_3(t)} \\ 0 & 0 & 0 & 1 \\ \frac{\rho\cos x_3(t)}{1-\rho^2\cos^2 x_3(t)} & 0 & 0 & \frac{-x_4(t)\rho^2\cos x_3(t)\sin x_3(t))}{1-\rho^2\cos^2 x_3(t)} \end{pmatrix} \tag{4.39}$$

$$\mathbf{B}(x_3(t)) = g(x_3(t)) \quad \mathbf{C} = \begin{pmatrix} 0 & 0 & 1 & 0 \\ 0 & 0 & 0 & 1 \end{pmatrix}.$$

Table 4.1 lists the nominal values of the parameters. They correspond to the laboratory version of the TORA system described in [BBC95]. As can be observed from the equations above, applying the procedures in Section 4.1 to obtain the TP canonical form (4.4) is extremely tedious and unlikely to be successful.

4.3.2 TP canonical model

We apply the TP model transformation of Chapter 3 to the qLPV model (4.38) to numerically extract the canonical form. First, we need to define the transformation space Ω. From the simulation results in the special issue of [Ber98] and also the works of [Tad01], [TTW98], [EOSR99], we observe that θ is always smaller than 0.85 rad. Also, according to the maximum allowable torque of $u = 0.1$ N, $\dot\theta$ will not be larger than 0.5 rad/sec. The control specifications in later chapters would include some additional information on Ω. Overall, with a little margin of overshooting for $\dot\theta$, we can set $\Omega = [-a, a] \times [-a, a]$, that is, $x_3(t) \in [-a, a]$ in unit rad and $x_4(t) \in [-a, a]$ in unit rad/sec, where $a = \frac{45}{180}\pi$. With Ω defined, we let the density of the discretization grid be 137×137 on $(x_3(t) \in [-a, a]) \times (x_4(t) \in [-a, a])$.

The result of the TP model transformation shows that TORA system (4.38) can be represented exactly in the TP canonical model form with minimum $5 \times 2 = 10$

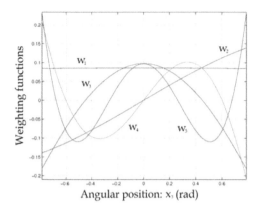

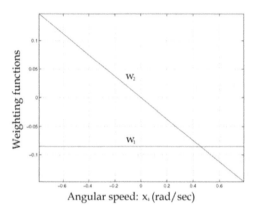

Figure 4.2: Weighting functions of the HOSVD-based canonical form.

LTI vertex models:

$$\dot{\mathbf{x}}(t) = \mathbf{S}(\mathbf{p}(t))\begin{pmatrix}\mathbf{x}(t)\\u(t)\end{pmatrix} = \sum_{i=1}^{5}\sum_{j=1}^{2} w_{1,i}(x_3(t))w_{2,j}(x_4(t))\left(\mathbf{A}_{i,j}\mathbf{x}(t) + \mathbf{B}_{i,j}u(t)\right). \qquad (4.40)$$

The resulting weighting functions are depicted in Figure 4.2. Note that the weighting functions here are orthogonal, instead of orthonormal, to each other. This is because the algorithm for generating the weighting functions, derived based on the results in Section 4.2 above, is aimed at producing orthonormal weighting functions when the grid density tends to a large value. For the moderate 137×137 density grid that we are using here, orthonormality is not yet reached. One can of course multiple appropriate scaling factors to the weighting functions to make them orthonormal, with corresponding reverse scaling to the LTI vertex models. Numerically checking between the original dynamics Equation (4.38) and TP canonical model (4.40) yields an error of less than 10^{-14}, which mainly comes from the numerical computation of HOSVD.

Chapter 5

Approximation and Complexity Trade-Off

This chapter presents the reasons and advantages of the approximation and complexity trade-off readily supported by the tensor product (TP) model transformation through the extraction of nonexact representation of quasi-linear parameter-varying (qLPV) models. Specifically, Sections 5.1 and 5.2 show mathematically that the set of functions possibly approximated to arbitrary degrees of accuracy by TP models with a *bounded* number of components lies nowhere dense in the set of continuous functions. This property hence necessitates a trade-off between the accuracy and complexity of the TP form for a certain class of modeling and control problems. The present content is based mostly on the works of [TBP02], [BKPH04], [TBP07], [BPK+07], [BYVKP03], [BYVK+02]. Again, readers who are less interested in the mathematical details of the theory may skip the first two sections and move to the examples given in Section 5.3.

5.1   TP model form of bounded order

For simplicity, we consider only bivariate TP models with only two parameter variables ($N = 2$). All definitions and statements can be carried over trivially to the multivariate case. Therefore, all results presented in the section are equally valid for any finite $N \in \mathbb{N}$ value. Further, without the loss of generality, we restrict the range of input to the unit interval, because any finite ranged intervals can be mapped into one another.

Definition 5.1. We refer to the matrix function of the following form as the TP form of order (I_1, I_2),

$$TP(z_1, z_2) = \mathbf{S}\begin{pmatrix} z_1 \\ z_2 \end{pmatrix} = \left(\sum_{i_1=1}^{I_1} \sum_{i_2=1}^{I_2} w_{1,i_1}(z_1) w_{2,i_2}(z_2) \mathbf{S}_{i_1,i_2} \right) \begin{pmatrix} z_1 \\ z_2 \end{pmatrix}, \qquad (5.1)$$

where the number of the component is bounded by I_1 in the first variable and by I_2 in the second variable.

As we noted above, all definitions and proofs can be generalized to the multivariate case, where the number of parameters z_n goes to N, namely, $n = 1, \ldots, N$. Let us denote the set of TP forms (5.1) defined in Definition 5.1 by

$$\mathbf{TP}_{(I_1,I_2)}\left([0, 1]^2\right), \qquad (5.2)$$

where the pair (I_1, I_2) is an upper bound for the number of components in the input space $[0, 1]^2$. For $p \in [1, \infty]$ we introduce the set

$$\mathbf{TP}^{(p)}_{(n_1,n_2)}\left([0, 1]^2\right), \qquad (5.3)$$

which is the subset of (5.2) and $L^p\left([0, 1]^2\right)$ as well. Therefore, the set (5.3) is equipped with the L^p-norm $\| \cdot \|_p$. An element of sets (5.3) and (5.2) is a TP model form of order (I_1, I_2). In general, TP model forms with finite bounds in each dimension are called TP model forms of bounded order.

5.2   The nowhere dense property

Moser [Mos99] has proven the nowhere denseness of Sugeno controllers [Sug85] in two steps. Our proof to be presented here is an extension of Moser's result, and its deduction technique is analogous to Moser's. First, we prove that a special function cannot be approximated arbitrarily well by TP model forms of bounded order regardless of what L^p-norm is chosen (cf. Lemma 4). It is then stated (see Theorem 5.1) that for every TP model form, t, and for arbitrary $\varepsilon > 0$, there exists a function ω in the ε-environment of s, which cannot be approximated with arbitrary accuracy by TP model forms of bounded order. This immediately implies that TP model forms of bounded order are nowhere dense in the space of continuous functions. The construction of ω in Theorem 5.1 is based on the special function of the lemma.

Now we point out that if there is one element of a matrix function TP that cannot be approximated arbitrarily well by the corresponding elements of bounded TP model forms, then this property can be carried over for the whole matrix function. Therefore in the next lemma, we investigate only one element of matrix function TP,

$t(z_1, z_2) = TP(z_1, z_2)_{[i,j]}$. That element is a linear function of (z_1, z_2) according to the TP model, where the coefficients depend on the $[i, j]$ elements of $\mathbf{S}_{i_1, i_2}$ matrices. For brevity, we refer to t as the TP model form hereafter, keeping in mind that it is an arbitrary element of the matrix function TP.

Lemma 4. Let $\omega : [0, 1] \rightarrow \mathbb{R}$ be a function of the form $\omega(z_1, z_2) = \alpha$ if $z_2 \geq z_1$, $\omega(z_1, z_2) \neq \alpha$ else, where $\alpha \in \mathbb{R}$. Then for each $p \in [1, \infty]$ and $I_1, I_2 \in \mathbb{N}$ there holds

$$\inf \left\{ \|\omega - t\|_p \mid t = TP_{[i,j]}, \ TP \in \mathbf{TP}^{(p)}_{(I_1, I_2)} \right\} > 0 \tag{5.4}$$

for an arbitrary (i, j) pair.

Proof. The proof proceeds indirectly. First, we suppose that opposing the assumption (5.4) there exists a sequence $\hat{t}_n$, $n \in \mathbb{N}$ of TP model forms, $t \in \left(\mathbf{TP}^{(p)}_{(I_1, I_2)} \right)_{[i,j]}$, which ensures norm-convergence to the function ω:

$$\lim_n \|\hat{t}_n - \omega\|_p = 0, \tag{5.5}$$

where $\hat{t}_n$, being TP model forms of order (I_1, I_2), have the form

$$\hat{t}_n(z_1, z_2) = \sum_{i_1=1}^{I_1} \sum_{i_2=1}^{I_2} w_{1,i_1}^{(n)}(z_1) w_{2,i_2}^{(n)}(z_2) \left(\widehat{\mathbf{S}}_{i_1, i_2}^{(n)} \right)_{[i,j]},$$

where $w_{1,i_1}^{(n)}(z_1), w_{2,i_2}^{(n)}(z_2)$ are basis functions and $\left(\widehat{\mathbf{S}}_{i_1, i_2}^{(n)} \right)_{[i,j]}$ are real numbers. If we modify properly TP model forms $\hat{t}_n$, then we obtain TP model forms t_n, according to the assumption (5.5) that approximate arbitrarily well functions having value 0 for $y \geq x$ and that differ from 0 for $y < x$ with respect to $\| \cdot \|_p$. Therefore, we introduce a sequence of TP model forms (t_n) $(n \in \mathbb{N})$ as

$$t_n(z_1, z_2) = \sum_{i_1=1}^{I_1} \sum_{i_2=1}^{I_2} w_{1,i_1}^{(n)}(z_1) w_{2,i_2}^{(n)}(z_2) \left(\mathbf{S}_{i_1, i_2}^{(n)} \right)_{[i,j]}, \tag{5.6}$$

where $\mathbf{S}_{[i,j]}^{(n)} = \widehat{\mathbf{S}}_{[i,j]}^{(n)} - \alpha$. Because norm-convergence implies pointwise convergence at almost everywhere from (5.5) and (5.6), we have

$$\lim_n \hat{t}_n = \omega - \alpha. \tag{5.7}$$

Let Ω' be the set where pointwise convergence holds.

Moser shows that there are points of the open four-dimensional unit hypercube $(0, 1)^4$

$$\left(z_{1,1}^*, z_{2,1}^* \right), \left(z_{1,1}^*, z_{2,2}^* \right), \text{ and } \left(z_{1,2}^*, z_{2,2}^* \right) \tag{5.8}$$

which satisfy the following inequalities (see also Figure 5.1)

$$z_{2,1}^* < z_{1,1}^* \leq z_{2,2}^* < z_{1,2}^* \tag{5.9}$$

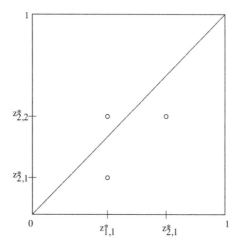

Figure 5.1: The points $(z_{1,1}^*, z_{2,1}^*)$, $(z_{1,1}^*, z_{2,2}^*)$, and $(z_{1,2}^*, z_{2,2}^*)$.

as well as lie in the set Ω and do not fall only in the set $[0, 1]^2/\Omega$ that has Lebesgue measure zero. These points are essential in the last step of the proof.

Now, we apply diagonal construction due to Cantor on the basis of Ω^*, which is a countable subset of Ω' lying dense in $[0, 1]^2$ as well. Therefore, we can write that $\{(z_{1,q}, z_{2,q}) | i \in \mathbb{N}\} = \Omega^* \subset \Omega$. Note that Ω^* also contains points in (5.8). For simplicity, let us denote $\Omega_{z_1}^* = z_{1,1}^*, z_{1,2}^*, \ldots$ and $\Omega_{z_2}^* = z_{2,1}^*, z_{2,2}^*, \ldots$.

Now, we require the sequence of appropriate elements of $\mathbf{S}$

$$\left(\mathbf{S}_{i_1, i_2}^{(q,n)}\right)_{[i,j]} \tag{5.10}$$

to converge to a real number as n tends to infinity. By the upper index (q, n) we denote the subsequence of the sequence denoted by upper index (n). Furthermore, we demand the sequence with upper index $(q+1, n)$ to be the subsequence of the sequence with upper index (q, n). Finally, from the above we may suppose that sequences $\left(w_{1,i_1}^{(n,n)}\right)_n$, $\left(w_{2,i_2}^{(n,n)}\right)_n$, and similarly $\left(\left(\mathbf{S}_{i_1,i_2}^{(n,n)}\right)_{[i,j]}\right)_n$ are convergent for all $(z_{1,q}, z_{2,q}) \in \Omega^*$ and all indices $1 \le i_1 \le I_1$, $1 \le i_2 \le I_2$.

It follows that the input-output functions of TP model forms $t_{n,n}$ converge to the input-output function of the TP model form

$$\tilde{t}(z_1, z_2) = \sum_{i,j} \tilde{w}_{1,i_1}(z_1)\tilde{w}_{2,i_2}(z_2)\left(\tilde{\mathbf{S}}_{i_1,i_2}\right)_{[i,j]}, \tag{5.11}$$

where

$$\lim_n w_{1,i_1}^{(n,n)} = \tilde{w}_{1,i_1}, \quad \lim_n w_{2,i_2}^{(n,n)} = \tilde{w}_{2,i_2},$$

$$\lim_n \left(\mathbf{S}_{i_1,i_2}^{(n,n)} \right)_{[i,j]} = \left(\tilde{\mathbf{S}}_{i_1,i_2} \right)_{[i,j]}.$$

Equation (5.11) is a tensor product of univariate functions with bounded order $J = I_1 \cdot I_2$, hence we can introduce induction similarly as in [Mos99].

Let us denote by $\mathbf{Ten}_J(D, Z_1 \times Z_2)$ those functions $f : Z_1 \times Z_2 \mapsto \mathbb{R}$ whose restrictions $f|_D$ can be represented as the tensor product of maximal $J \in \mathbb{N}$ basis functions for each dimension, that is, for an $f \in \mathbf{Ten}_n(D, Z_1 \times Z_2)$ iff $f(z_1, z_2) = \sum_{i=1}^{J} \phi_i(z_1) \psi_i(z_2)$ holds for all $(z_1, z_2) \in D$, where $\phi_1, \dots, \phi_J : Z_1 \mapsto \mathbb{R}$ and $\psi_1, \dots, \psi_J : Z_2 \mapsto \mathbb{R}$.

In order to deduce a contradiction from the assumption that ω can be approximated arbitrarily well by TP model forms, we proceed to show that in fact the limit input-output function of TP model forms (5.11) is an element of $\mathbf{Ten}_1(\Omega^* \cap [\gamma, 1]^2, [0, 1]^2)$, for some $0 < \gamma < z_{2,1}^*$. Then from the fact that the function $\tilde{t}(z_1, z_2)$ equals by assumption the function $\omega - \alpha$ on the domain Ω^* and the construction that the points (5.8) are elements of $\Omega^* \cap [\gamma, 1]^2$, a contradiction can immediately be derived analogously as in [Mos99].

Since $\tilde{t} \equiv \omega - \alpha$ on Ω^*, there must be an index $i_0 \in \{1, \dots, J\}$ for which $\phi_{i_0} \not\equiv 0$ on $\Omega_{z_2}^* \cap (0, z_{2,1}^*/4)$.

Without loss of generality let $i_0 = J$. Thus, there is an element

$$z_{1,0} \in \Omega_{z_1}^* \cap (0, z_{2,1}^*/4), \tag{5.12}$$

such that $\phi_J(z_{1,0}) \neq 0$. Consequently, for $z_2 \in \Omega_{z_2}^* \cap [z_{1,0}, 1]$ we obtain

$$\psi_J(z_2) = -\sum_{i=1}^{J-1} \psi_i(z_2) \frac{\phi_i(z_{1,0})}{\phi_J(z_{1,0})},$$

as the function $\tilde{t}$ vanishes on $\{(z_1, z_2) \in \Omega^* | z_2 \geq z_1\}$. Hence, for all pairs $(z_1, z_2) \in \Omega^*$ with $z_1, z_2 \geq z_{2,1}^*/4$ we get

$$\tilde{t}(z_1, z_2) = \sum_{i=1}^{J-1} \left(\phi_i(z_1) - \phi_J(z_1) \cdot \frac{\phi_i(z_{1,0})}{\phi_J(z_{1,0})} \right) \psi_i(z_2)$$

showing that

$$\tilde{t}(z_1, z_2) \in \mathbf{Ten}_{J-1} \left(\Omega^* \cap [z_{2,1}^*/4, 1]^2, [0, 1]^2 \right). \tag{5.13}$$

In order to apply induction we set

$$\gamma_k = (1 - (3/4)^{J-k}) z_{2,1}^*$$

for $1 \leq k \leq J$. Analogously as we have deduced (5.13) from (5.12),

$$\tilde{t}(z_1, z_2) \in \mathbf{Ten}_{k-1} \left(\Omega^* \cap [\gamma_{k-1}, 1]^2, [0, 1]^2 \right)$$

can be deduced from

$$\tilde{t}(z_1, z_2) \in \mathbf{Ten}_k \left(\Omega^* \cap [\gamma_k, 1]^2, [0, 1]^2 \right)$$

for $(2 \leq k \leq J)$. Applying induction we obtain

$$\tilde{t}(z_1, z_2) \in \mathbf{Ten}_1 \left(\Omega^* \cap [\gamma_1, 1]^2, [0, 1]^2 \right). \tag{5.14}$$

From formula (5.14) we conclude that there are functions ϕ and $\psi : [0, 1] \mapsto \mathbb{R}$ such that

$$\tilde{t}(z_1, z_2) = \omega(z_1, z_2) - \alpha = \phi(z_1)\psi(z_2) \tag{5.15}$$

for $(z_1, z_2) \in \Omega^* \cap (\gamma_1, 1]^2$.

As $\gamma_k < z_{2,1}^*$ for all $1 \leq k \leq J$, the points (5.8) are elements of the set $\Omega^* \cap (\gamma_1, 1]^2$.

This and Equation (5.15) lead to a contradiction as on the one hand we have $\phi(z_{1,1}^*)\psi(z_{2,1}^*) > 0$ and $\phi(z_{1,2}^*)\psi(z_{2,2}^*) > 0$, hence

$$\phi(z_{1,1}^*) \neq 0 \quad \text{and} \quad \psi(z_{2,2}^*) \neq 0,$$

while on the other hand, by the definition of ω, we obtain $\phi(z_{1,1}^*)\psi(z_{2,2}^*) = 0$. □

Lemma 4 exemplifies a function that cannot be approximated arbitrarily well by TP model forms of order (I_1, I_2) with respect to the L_p norm. The function ω that is approximated above is thus not an element of $cl\left(\mathbf{TP}_{(I_1, I_2)}^{(p)}\left([0, 1]^2\right)\right)$, that is, the closure (w.r.t. the L_p norm) of the subset $\mathbf{TP}_{(I_1, I_2)}^{(p)}\left([0, 1]^2\right)$ of $L_p\left([0, 1]^2\right)$.

While Lemma 4 only states the existence of such a function, the next theorem states that $L_p\left([0, 1]^2\right) \setminus cl\left(\mathbf{TP}_{(I_1, I_2)}^{(p)}\left([0, 1]^2\right)\right)$ lies dense in $L_p\left([0, 1]^2\right)$. The idea is to show that in each neighborhood of an arbitrary function one can find a function which cannot be approximated arbitrarily well by TP forms of bounded order. These neighboring functions are constructed by means of Lemma 4. The proof basically exploits the special form of ω, and it can be trivially obtained from the proof of the nowhere denseness of Sugeno controllers [Mos99]. It is thus omitted here.

Theorem 5.1. *To each $p \in [1, \infty]$, $\varepsilon > 0$ and $t = TP_{[i,j]}, TP \in \mathbf{TP}_{(I_1, I_2)}^{(p)}\left([0, 1]^2\right)$, $I_1, I_2 \in \mathbb{N}$, there is a continuous function*

$$\omega \in L^p([0, 1]^2) \setminus cl\left(\mathbf{TP}_{(I_1, I_2)}^{(p)}\left([0, 1]^2\right)\right)$$

fulfilling $\|\omega - t\|_p < \varepsilon$.

As Theorem 5.1 guarantees that to each $\varepsilon > 0$ and $t = TP_{[i,j]}$, $TP \in \mathbf{T}_{(l_1,l_2)}^{(p)}\left([0,1]^2\right)$ there is a function $\omega \in L^p([0,1]^2)\backslash cl\left(\mathbf{TP}_{(l_1,l_2)}^{(p)}\left([0,1]^2\right)\right)$ with $\|\omega - t\|_p < \varepsilon$, or, equivalently, there is no inner point of the set $\mathbf{TP}_{(l_1,l_2)}^{(p)}\left([0,1]^2\right)$, because every ε-environment of an arbitrary element of the set contains functions not included in the closure of the set. Thus, we immediately obtain that $\mathbf{TP}_{(l_1,l_2)}^{(p)}$ is nowhere dense in $L^p([0,1]^2)$.

5.3 Trade-off examples

In this section we present two examples derived from the second-order mechanical system of Figure 5.2. The first is a nonlinear mass-spring-damper system that can be exactly represented by a TP model. Derivation of the exact TP representation in this case can be obtained analytically (see [TIW96]), as well as computationally using the TP model transformation introduced above. The second example adds to the mechanical system of the first example a nonlinear term having the same properties as ω in Lemma 4, which makes the exact TP model representation impossible regardless of the number of weighting functions being used. The example thus serves as a synthetical case study to show that the TP model transformation has the ability to deal with trade-off processing that in most cases is not possible analytically. The ability to trade off is important here because, as Part II of the book shows, the complexity of the linear matrix inequality (LMI)-based design increases exponentially (except for a few special cases) with the number of the components of the resulting TP model. This fact has a significant implication on the maximum number of components of the TP model one should accept, given the detrimental effects due to the reduced accuracy of a lesser representation. With the fact that the TP model and TP canonical form are on the same level as far as trading-off between complexity and accuracy is concerned, in the following we will present the trade-off results using the TP canonical model.

5.3.1 A mass-spring-damper system

The dynamical equation of the mechanical system of Figure 5.2 is given by:

$$m\ddot{x}(t) + g(x(t), \dot{x}(t)) + k(x(t)) = \phi(\dot{x}(t))u(t), \qquad (5.16)$$

where m is the mass, $u(t)$ is the input force to the mass, and $\phi(\dot{x}(t))$ is the nonlinear multiplicative coefficient of the input term. The function $g(x, \dot{x})$ is the nonlinear force of the damper and $k(x)$ is the nonlinear restoring force of the spring. The directions of these two forces for the case of positive $x(t)$ and $\dot{x}(t)$ are as shown in Figure 5.2. Both forces may also embed uncertainties in their descriptions. Here, we set $g(x(t), \dot{x}(t)) =$

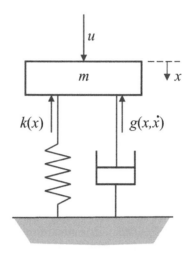

Figure 5.2: Mass-spring-damper system.

$d(c_1x(t)+c_2\dot{x}^3(t)), k(x(t)) = c_3x(t)+c_4x^3(t)$, and $\phi(\dot{x}(t)) = 1+c_5\dot{x}^3(t)$, with $x \in [-a,a]$, $\dot{x}(t) \in [-b,b]$, and $a,b > 0$.

We consider the case where $m = 1, d = 1, c_1 = 0.01, c_2 = 0.1, c_3 = 0.01$, $c_4 = 0.67, c_5 = 0, a = 1.5$, and $b = 1.5$. Equation (5.16) then becomes:

$$\ddot{x}(t) = -0.1\dot{x}^3(t) - 0.02x(t) - 0.67x^3(t) + u(t). \tag{5.17}$$

The nonlinear terms are $-0.1\dot{x}^3(t)$ and $-0.67x^3(t)$. The state space model is thus:

$$\dot{\mathbf{x}}(t) = \begin{pmatrix} -0.1x_1^2(t) & -0.02 - 0.67x_2^2(t) \\ 1 & 0 \end{pmatrix}\mathbf{x}(t) + \begin{pmatrix} 1 \\ 0 \end{pmatrix}u(t), \tag{5.18}$$

or

$$\dot{\mathbf{x}}(t) = \mathbf{A}(x_1(t), x_2(t))\mathbf{x}(t) + \mathbf{B}u(t), \tag{5.19}$$

where

$$\mathbf{x}(t) = \begin{pmatrix} \dot{x}(t) \\ x(t) \end{pmatrix}; \quad \mathbf{A}(x_1(t), x_2(t)) = \begin{pmatrix} -0.1x_1^2(t) & -0.02 - 0.67x_2^2(t) \\ 1 & 0 \end{pmatrix}; \quad \mathbf{B} = \begin{pmatrix} 1 \\ 0 \end{pmatrix}. \tag{5.20}$$

Upon the application of TP model transformation, we readily obtain the following exact TP canonical model of (5.18) as:

$$\dot{\mathbf{x}}(t) = \mathcal{A} \underset{n=1}{\overset{2}{\boxtimes}} \mathbf{w}_n(x_n(t))\mathbf{x}(t) + \mathbf{B}u(t). \tag{5.21}$$

The weighting functions of the TP model $\mathbf{w}_1(x_1(t))$ and $\mathbf{w}_2(x_2(t))$, with $x_1 = \dot{x}$ and $x_2 = x$, are depicted in Figure 5.3. In this case, exact representation of (5.18) by

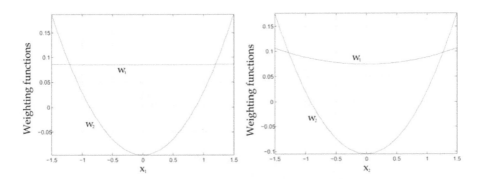

Figure 5.3: Weighting functions from TP model transformation of system (5.18).

the canonical model is possible with two weighting functions in each dimension, as depicted in the figure. Note that the weighting functions are orthogonal, instead of orthonormal, to each other, as explained in Section 4.3.2. The representation (5.21) can also be derived analytically from (5.18), something the readers may want to try.

5.3.2 A mass-spring-damper system with nonlinear term

Let us consider the dynamic system (5.16) with an extra term $f(x(t), \dot{x}(t))\dot{x}$ added to the nonlinear term $g(x(t), \dot{x}(t))$ as

$$g(x(t), \dot{x}(t)) = d(c_1 x(t) + f(x(t), \dot{x}(t))\dot{x}(t) + c_2 \dot{x}^3(t))$$

where

$$f(x_1(t), x_2(t)) = \begin{cases} 0.01 & \text{if } x_2(t) \geq x_1(t) \\ 0.01 + (x_1(t) - x_2(t)) & \text{if } x_2(t) < x_1(t) \end{cases}$$

is a continuous function constructed from the ω in Lemma 4 and depicted in Figure 5.4. Here, the parameters are set as follows: $c_1 = 0.01$, $c_2 = 0$, $c_3 = 0.01$, $c_4 = 0$, $c_5 = 0$. All other parameters remain unchanged. Then (5.16) becomes

$$\ddot{x}(t) = -f(x(t), \dot{x}(t))\dot{x}(t) - 0.02x(t) + u(t),$$

and in state space form

$$\dot{\mathbf{x}}(t) = \begin{bmatrix} -f(x_1(t), x_2(t)) & -0.02 \\ 1 & 0 \end{bmatrix} \mathbf{x}(t) + \begin{pmatrix} 1 \\ 0 \end{pmatrix} u(t). \tag{5.22}$$

Let us now study the characteristics of the TP model transformation in terms of accuracy and complexity. A TP model, or TP canonical model, of (5.22) requires infinity terms for exact representation. Analytical derivation would be impossible in

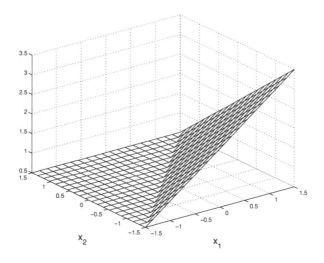

Figure 5.4: The plot of function $f(x_1, x_2)$ over the region $[-1.5, 1.5] \times [-1.5, 1.5]$.

this case. On the other hand, the TP model transformation can produce an inexact representation with a finite number of components because it is a numerical method. Figure 5.5 shows the result of using 2×2 weighting functions on the dimensions of the state vector. The maximum error between the resulting representation in this case with the original model of Equation (5.22) is 7.8. In order to improve the approximation accuracy, we can increase the number of the weighting functions to 3×2 as in Figure 5.6. The maximum error is then decreased to 4.0425. Along the same line we can generate a representation with 3×3 weighting functions as depicted in Figure 5.7 that yields a maximum error of 2.6152. Again, as mentioned, the weighting functions in the figures are all orthogonal, instead of orthonormal, to each other.

In reality, the maximum number of components can be set according to the computational loading we can afford versus the approximation accuracy we would like to achieve. Both factors will have ramifications to the ensuing controller design process later. Table 5.1 shows the decrease of the approximation error with an increasing number of weighting functions for possible trade-off consideration between the number of components in the TP representation and its corresponding accuracy. Note that such results would be very difficult to obtain via analytical derivations.

5.4   Trade-off study on the TORA example

We continue with the translational oscillations with an eccentric rotational proof mass actuator (TORA) system example in Section 4.3. A detailed study of this system can be found in [PBH10]. The goal is to see how we can perform a trade-off

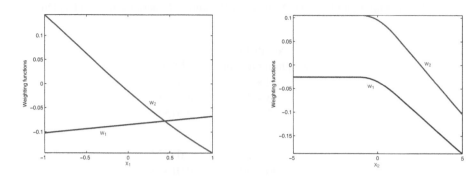

Figure 5.5: 2×2 weighting functions for TP representation of (5.22).

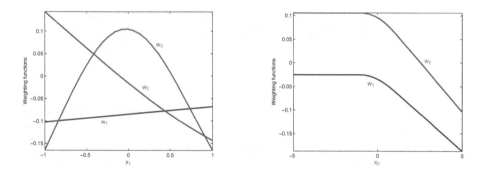

Figure 5.6: 3×2 weighting functions for TP representation of (5.22).

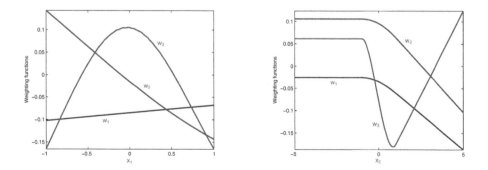

Figure 5.7: 3×3 weighting functions for TP representation of (5.22).

Table 5.1: Maximum Error vs. Number of Weighting Functions in x_1 and x_2 Dimensions

Number of weighting functions	Error
2×2	7.8
3×2	4.0425
3×3	2.6152
$\vdots$	$\vdots$
20×20	0.02838
30×30	0.01476
40×40	0.00895
50×50	0.00555

between the complexity and the accuracy of the TORA TP model. The previous results indicate that a TP canonical model would require a minimal number of 5×2 linear time-invariant (LTI) components to exactly represent the TORA system. Here, we will discard the singular values sequentially and check the accuracy of the resulting model.

Figure 5.8 shows the result of discarding the smallest singular value on the dimension of x_3. Figure 5.9 shows the same of discarding the two smallest singular values and Figure 5.10 that of discarding the three smallest singular values. No change in the dimension x_4 is made. Again, the weighting functions are orthogonal as explained. The maximum errors caused by the discarded singular values, checked numerically over a huge number of points in Ω, are 0.0005, 0.0017, and 0.0465, respectively, for the three cases. The results clearly indicate the trade-off between model complexity and accuracy.

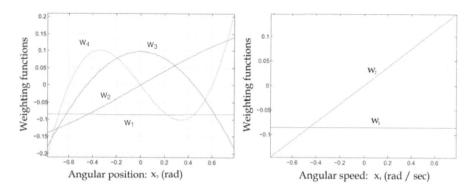

Figure 5.8: Weighting functions of the TP canonical model of the TORA system keeping four singular values in the x_3 dimension.

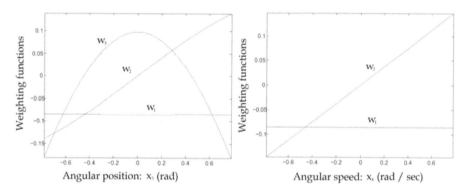

Figure 5.9: Weighting functions of the TP canonical model of the TORA system keeping three singular values in the x_3 dimension.

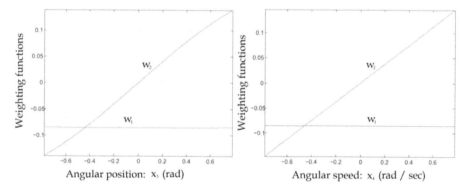

Figure 5.10: Weighting functions of the TP canonical model of the TORA system keeping two singular values in the x_3 dimension.

Chapter 6

TP Model Convexity Incorporation

As will be seen in Part II of this book, tensor product (TP) models with convexity offer great advantages in the process of control system design. It is therefore highly desirable to further transform the higher-order singular value decomposition (HOSVD)-based canonical form of quasi-linear parameter-varying (qLPV) models to become TP model incorporating various convexities. The convexity induced will in turn define the type of convex hull produced by the linear time-invariant (LTI) vertex systems of the resulting TP model, and yield strong influence on the control design feasibility and performance. This chapter serves to present a systematic way of generating different types of convex TP models.

It would be ideal if we can directly define the vertices of the convex hull by setting the LTI vertex systems and then computing the weighting functions accordingly to form a tightly bounded convex hull. This is, however, not possible because the vertex systems and the weighting functions are coupled together to effectuate an acceptable approximation of the qLPV model. One cannot arbitrarily change one without changing the other, lest we introduce a large error and destroy the accuracy of the TP model.

Instead, we proceed on the following path. As it has been established that the HOSVD-based canonical form of a qLPV model can be numerically reconstructed via the TP model transformation of Chapter 3, we can thus focus on how to incorporate different types of convex TP models during the process of TP model transformation. Specifically, recall that in Step 2 of the transformation procedures, after we have obtained

$$\mathcal{S}^D = \mathcal{S} \underset{n=1}{\overset{N}{\boxtimes}} \mathbf{U}_n,$$

we also introduce the possibility of transforming $\mathbf{U}_n$ to become $\bar{\mathbf{U}}_n$ through a transformation $\mathbf{T}_n$, $\bar{\mathbf{U}}_n \mathbf{T}_n = \mathbf{U}_n$, resulting in

$$\mathcal{S}^D = \bar{\mathcal{S}} \underset{n=1}{\overset{N}{\boxtimes}} \bar{\mathbf{U}}_n, \tag{6.1}$$

where $\bar{\mathcal{S}} = \left(\mathcal{S} \underset{n=1}{\overset{N}{\boxtimes}} \mathbf{T}_n \right)$. The quantity $\bar{\mathcal{S}}$ can also be obtained directly from the discretized samples $\mathcal{S}^D$ and the desired $\bar{\mathbf{U}}_n$ via

$$\bar{\mathcal{S}} = \mathcal{S}^D \underset{n=1}{\overset{N}{\boxtimes}} \bar{\mathbf{U}}_n^+, \tag{6.2}$$

where "+" means pseudo-inverse. Note that the transformation $\mathbf{T}_n$ must be full rank so that $\bar{\mathbf{U}}_n$ is equal in rank to $\mathbf{U}_n$ and ensures the same approximation accuracy of the TP models with (6.1) and (6.2). An incorrect selection may decrease the rank of $\bar{\mathcal{S}}$ and degrade the approximation property of the eventual TP model.

The idea here is that, while the matrix $\mathbf{U}_n$ contains orthonormal columns which may not produce any convexity, we can design proper transformations $\mathbf{T}_n$ to inject various types of convexity into $\bar{\mathbf{U}}_n$ and take (6.1) as the outcome of Step 2. Then in Step 3 of the TP model transformation, the identified nth-dimension weighting functions evaluated at discretized values of p_n, which is given by $\bar{\mathbf{U}}_n$, will inherit the same convexity condition. Then in Step 3 of the TP model transformation, $\bar{\mathbf{U}}_n$ is identified as the nth-dimension weighting functions evaluated at discretized values of p_n. We thus incorporate the same convexity condition to the discretized weighting functions. The question is, upon interpolating these discretized weighting functions to become continuous ones, would the convexity condition hold for the in-between points? As mentioned before, should linear interpolation be adopted to join the discretized values in the formation of the continuous weighting functions, the same convexity will be satisfied in-between. If another type of interpolation is adopted, in-between convexity will not be guaranteed. This latter case gives rise to the quasi-convexity characterization introduced in Section 6.1.

6.1   TP model convexity

In this section, we pick up what Chapter 2 has started and define the different types of convexities for matrix $\mathbf{U}_n$. We note that in the TP model transformation procedures of Section 3.2, matrix $\mathbf{U}_n$ is identified as the discretized weighting functions, which is then interpolated to yield the continuous weighting function $\tilde{\mathbf{U}}_n$ in Section 3.3. The convexity property of $\mathbf{U}_n$ can thus be readily transferred from $\mathbf{U}_n$ to the corresponding

TP models. For simplified notation, we will drop the subscript "*n*" from $\mathbf{U}_n$ and just use $\mathbf{U}$ when the meaning is clear.

Definition 6.1 (Sum normalization [SN] type matrix $\mathbf{U}$). A real matrix $\mathbf{U}$ is of the SN type if the sum of all elements in each row is 1.

Definition 6.2 (Nonnegativeness [NN] type matrix $\mathbf{U}$). A real matrix $\mathbf{U}$ is of the NN type if it does not have any negative element.

Note that if we discretize the weighting functions of an SN type TP model (see Definition 2.6), then we obtain an SN type matrix $\mathbf{U}$. Similarly, if we discretize the weighting functions of a NN type TP model (see Definition 2.5), we obtain a NN type matrix $\mathbf{U}$. On the other hand, however, an SN or NN type matrix $\mathbf{U}$ does not necessarily produce an SN or NN type TP model. That will depend on how we interpolate the discretized weighting function defined by $\mathbf{U}$ to form continuous weighting functions in Step 3 of the TP model transformation.

If we adopt linear interpolation, then it is true that SN and NN type matrices $\mathbf{U}$ will ensure an SN and NN type TP model, respectively. However, if we define the continuous weighting functions by other interpolation techniques, we will have to check whether the resulting continuous weighting functions satisfy the SN and NN condition in-between discretization points. If not, or we do not know how the in-between points behave, we say that the TP model is of quasi-SN (qSN) or quasi-NN (qNN) type respectively.

Recall that we have defined a convex TP model to be a TP model of both SN and NN type (Definition 2.7). As will be seen later in this chapter, a TP model extracted from HOSVD-based procedures can always be made SN and NN, and hence convex. SN and NN are sort of the basic properties of a TP model. In the following, we define other convexity types that may not be possibly incorporated in a TP model.

Definition 6.3 (Normality [NO] type matrix $\mathbf{U}$). A real matrix $\mathbf{U}$ is of NO type if it is SN and NN, and the largest value of each column is 1.

Definition 6.4 (Close-to-normality [CNO] type matrix $\mathbf{U}$). A real matrix $\mathbf{U}$ is of CNO type if it is SN and NN, and the largest value of each column is 1 or close to 1.

Definition 6.5 (Relaxed normality [RNO] type matrix $\mathbf{U}$). A real matrix $\mathbf{U}$ is of RNO type if it is SN and NN, and the largest values of all columns are the same. This value is always between 0 and 1.

Definition 6.6 (Inverse normality [INO] type matrix $\mathbf{U}$). A real matrix $\mathbf{U}$ is of INO type if it is SN and NN, and the smallest value of all columns is 0.

Assuming linear interpolation of $\mathbf{U}$ is adopted to form the continuous weighting functions of the TP model, we have the following:

Definition 6.7 (NO type TP model). A TP model is of NO type if it is convex and the maximum value of all weighting functions belonging to the same dimension is 1.

Definition 6.8 (CNO type TP model). A TP model is of CNO type if it is convex and the maximum value of all weighting functions belonging to the same dimension is 1, or close to 1.

Definition 6.9 (RNO type TP model). A TP model is of RNO type if it is convex and the maximum values of all weighting functions belonging to the same dimension are the same. This value is always between 0 and 1.

Definition 6.10 (INO type TP model). A TP model is of INO type if it is convex and the smallest values of all weighting functions are 0.

We can also define a TP model satisfying the conditions of both INO and RNO types as an INO-RNO type, or just simply IRNO type.

The geometric implication of the weighting function convexities or the TP model convexity types on the resultant convex hull of $S(p)$ can be described as follows. Consider $S(p)$ in the TP form (2.9) being mapped from Ω to its range space. As the linear combination in (2.9) is convex, the vertices of the convex hull of $S(p)$ are hence given by the vertex LTI systems $S_{i_1,...,i_N}$. A NO type TP model, with NO weighting functions in all dimensions, will yield a tighter convex hull than the CNO, RNO, and INO types. Specifically in this case, $S(p)$ will be equal to each of the vertices, $S_{i_1,...,i_N}$, at appropriate values of p in Ω. This occurs when p_n happens to be at such value that the $\bar{i}_n$th weighting function of the nth dimension attains the value of 1 for $n = 1,...,N$. As NO weighting functions are also SN and NN, this implies zero value for all other weighting functions. The TP form of $S(p)$ thus dictates that $S(p) = S_{\bar{i}_1,...,\bar{i}_N}$. As $\bar{i}_n$ goes from 1 to I_n for $n = 1,...,N$, the mapped image of $S(p)$ in the range space will visit each of the vertex LTI systems $S_{i_1,...,i_N}$. This is illustrated in case (a) of Figure 6.1.

For a CNO type TP model, not all weighting functions can acquire a value of 1 for $p_n \in [a_n, b_n]$. For these weighting functions, the maximum they can acquire are close to 1 values, with other weighting functions in the same dimension taking on small but nonzero values. From the TP form of (2.9), this means that the resulting $S(p)$ will be a linear combination of at least two or more of the associating vertex LTI systems. On the other hand, it is still possible for a CNO type TP model that some weighting functions acquire the values of 1 for $p_n \in [a_n, b_n], n = 1,...,N$, leading to $S(p)$ being exactly mapped into some of the vertex systems. The situation is illustrated in case (b) of Figure 6.1, showing that the mapped image of $S(p)$ deviates from some $S_{i_1,...,i_N}$ while being exactly equal to some others. In this case, the "degree" of deviations, if they occur, can vary from vertex to vertex. This is because CNO convexity does not require the maximum values of the weighting functions to be the same, only that they are 1, or close to 1. Hence, the contribution by the dominant vertex in (2.9) varies from vertex to vertex and results in varying "degrees" of deviation.

Case (c) of Figure 6.1 shows the RNO case, where we require the maximum value of all weighting functions in the same dimension to have the same close to 1 value. Compared to the CNO case, we see that $S(p)$ will not pass through any of the vertex LTI systems. This is because no weighting function can acquire the value 1 in the RNO case, and so $S(p)$ will surely be given by the linear combination of at least two vertices. Moreover, the "degree" of deviations of the mapped image of $S(p)$ from vertex to vertex are more uniform in the RNO case. This is because the contribution to $S(p)$ by the dominant vertex is more uniform over all the vertices as a result of the same close to 1 value attained in the corresponding weighting functions.

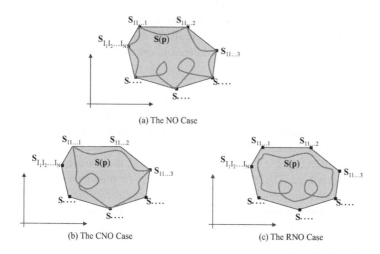

(a) The NO Case

(b) The CNO Case   (c) The RNO Case

Figure 6.1: Convex hull of $S(p)$ under different TP model convexity conditions.

Finally, the INO TP model case just ensures that some vertices will not contribute to $S(p)$ at all at specific values of $p \in \Omega$ because their corresponding weighting functions are of zero value. Note the NO type is a special case of the INO type—when $S(p)$ is mapped into a specific vertex, none of the other vertices are contributing.

6.2 Incorporation of convexity conditions

This section introduces the techniques to incorporate various convexity conditions of SN, NN, NO, CNO, RNO and INO. The content is based on the works of [Yam97], [YYB03], [Bar06a], [YBY99], [BY00] for the SN, NN, and NO cases, [YBY99], [BY00] for the SN, NN and NO/CNO case, and [YYB03], [Bar06a] for the RNO and INO cases. Let us first revisit the context for the problem at hand. Recall that after HOSVD of Theorem 3.2, we obtain (3.5)

$$\mathcal{A} = \mathcal{S} \times_1 \mathbf{U}_1 \times_2 \mathbf{U}_2 \times_3 \cdots \times_N \mathbf{U}_N = \mathcal{S} \overset{N}{\underset{n=1}{\boxtimes}} \mathbf{U}_n, \tag{6.3}$$

where $\mathbf{U}_n$, $n = 1, \ldots, N$ is an orthogonal and hence, a square matrix. Then, in the ensuing compact higher-order singular value decomposition (CHOSVD) or reduced higher-order singular value decomposition (RHOSVD) process, as the case may be, we will end up keeping a certain number of columns of $\mathbf{U}_n$ upon discarding some singular values to acquire the desired approximation property. Generally, the matrix with the retained columns will not inherit any convexity condition. This is demonstrated in the translational oscillations with an eccentric rotational proof mass actuator (TORA) example in Chapter 5. For all the cases considered, the weighting functions as generated from the retained columns are not SN nor NN, let alone others.

For simplicity, we will denote the orthogonal matrix $\mathbf{U}_n$ as matrix $\mathbf{U}$ (without the subscript) of size $(n \times n)$ here. Also, let $\mathbf{U} = [\mathbf{U}^{(r)} | \mathbf{U}^{(d)}]$, where $\mathbf{U}^{(r)}$ and $\mathbf{U}^{(d)}$ contain, respectively, the retained columns and the discarded columns. It is the goal here to design special transformation $\mathbf{T}$ to convert the $\mathbf{U}^{(r)}$ to become, say, $\bar{\mathbf{U}}^{(r)}$ that is SN, NN, NO/CNO, RNO, or INO, as the case may be. Note that we are working with an orthogonal matrix here. The conversion algorithms to be presented below will work just the same as for the orthonormal one that we encounter in the HOSVD process. Moreover, in previous formulations, the transformation $\mathbf{T}$ to inject convexity is to be multiplied in the form $\bar{\mathbf{U}}^{(r)}\mathbf{T} = \mathbf{U}^{(r)}$, with the corresponding quantity $\bar{\mathcal{S}}$ given by (6.1) and (6.2). Here, we allow the flexibility that the transformation can be multiplied in the form $\bar{\mathbf{U}}^{(r)} = \mathbf{U}^{(r)}\mathbf{T}$, in order to more conveniently inject the convexity. In this case, the quantity $\bar{\mathcal{S}}$ should be obtained by appropriately modifying (6.2).

6.2.1   Incorporating the SN condition

Let us define the following notation: sum($\mathbf{F}$) denotes the column vector obtained by summing over the rows of matrix $\mathbf{F}$. Now we use the matrix $\mathbf{U}$ to state the following theorem.

Theorem 6.1. *Let* $\mathbf{U}$ *be an* $n \times n$ *orthogonal matrix partitioned as* $\mathbf{U} = [\mathbf{U}^{(r)} | \mathbf{U}^{(d)}]$, *where* $\mathbf{U}^{(r)}$ *and* $\mathbf{U}^{(d)}$ *have, respectively,* $n^{(r)}$ *and* $n^{(d)} = n - n^{(r)}$ *columns. Let* $\mathbf{T}^{SN}$ *be any* $n^{(r)} \times n^{(r)}$ *transformation matrix satisfying the constraint* sum($\mathbf{T}^{SN}$) = sum(($\mathbf{U}^{(r)})^T$). *Then*

if sum(($\mathbf{U}^{(r)})^T$) = $(0)_{n^{(d)} \times 1}$, *the* $n \times n^{(r)}$ *matrix* $\bar{\mathbf{U}}$ *satisfies SN, with*

$$\bar{\mathbf{U}} = \mathbf{U}^{(r)}\mathbf{T}^{SN}, \tag{6.4}$$

and if $\text{sum}((\mathbf{U}^{(r)})^T) \neq (0)_{n^{(d)} \times 1}$, *the* $n \times (n^{(r)} + 1)$ *matrix* $\bar{\mathbf{U}}$ *satisfies SN, with*

$$\bar{\mathbf{U}} = [\mathbf{U}^{(r)} \mid \mathbf{U}^{(d)} \text{sum}((\mathbf{U}^{(d)})^T)] \begin{bmatrix} \mathbf{T}^{SN} & \mathbf{0}_{n^{(r)} \times 1} \\ \mathbf{0}_{1 \times n^{(r)}} & 1 \end{bmatrix}. \tag{6.5}$$

Theorem 6.1 characterizes the extra column needed, if at all, to supplement the matrix $\mathbf{U}^{(r)}$ in order to satisfy the SN condition. The proof can be found in [Yam97] and is omitted here. Note that the theorem does not require $\mathbf{T}^{SN}$ to be invertible. However, as mentioned before, an invertible $\mathbf{T}^{SN}$ does allow us to maintain the same approximation performance of $\mathbf{U}^{(r)}$ while using $\bar{\mathbf{U}}$. We present the following algorithm to construct $\mathbf{T}^{SN}$: if $\text{sum}(\mathbf{U}^{(r)^T})$ does not contain zero elements, form matrix $\mathbf{T}^{SN}$ as

$$\mathbf{T}^{SN} = \text{diag}[\text{sum}(\mathbf{U}^{(r)^T})],$$

and if $\text{sum}(\mathbf{U}^{(r)^T})$ contains zero element(s), form $\mathbf{T}^{SN}$ as

$$\mathbf{T}^{SN} = \mathbf{I}_{n^{(r)} \times n^{(r)}} + \left[\mathbf{0}_{n^{(r)} \times (\widetilde{n}-1)} | \text{sum}(\mathbf{U}^{(r)^T}) - \mathbf{1}_{n^{(r)} \times 1} | \mathbf{0}_{n^{(r)} \times (n^{(r)} - \widetilde{n})} \right]$$

and we need to ensure that the $\widetilde{n}$th entry of $\text{sum}(\mathbf{U}^{(r)^T})$ is nonzero. After $\mathbf{T}^{SN}$ is obtained, we then have

$$\begin{cases} \text{For } \text{sum}(\mathbf{U}^{(r)^T}) = \mathbf{0}_{(n-n^{(r)}) \times 1}, \\ \bar{\mathbf{U}} = \mathbf{U}^{(r)} \mathbf{T}^{SN} \\ \\ \text{For } \text{sum}(\mathbf{U}^{(r)^T}) \neq \mathbf{0}_{(n-n^{(r)}) \times 1}, \\ \bar{\mathbf{U}} = \left[\mathbf{U}^{(r)} | \mathbf{U}^{(d)} \text{sum}(\mathbf{U}^{(r)^T}) \right] \begin{bmatrix} \mathbf{T}^{SN} & \mathbf{0}_{n^{(r)} \times 1} \\ \mathbf{0}_{1 \times n^{(r)}} & 1 \end{bmatrix} \end{cases} \tag{6.6}$$

The matrix $\bar{\mathbf{U}}$ is now of the SN type. It has either $n^{(r)}$ or $n^{(r)} + 1$ columns, depending on the case. Note that it is always possible to incorporate the SN condition to the matrix $\mathbf{U}^{(r)}$.

6.2.2 Incorporating the NN condition

The SN matrix $\bar{\mathbf{U}}$ as resulted above may contain negative elements and hence does not satisfy the NN condition. The following gives a set of procedures to generate from an SN matrix another matrix with the same dimension satisfying both SN and NN conditions. Let $\bar{\mathbf{U}}$ be an $(n \times m)$ matrix satisfying SN, and we proceed to

1. Look for the minimum element, $\min_{s,t}(\bar{\mathbf{U}}_{s,t})$ of $\bar{\mathbf{U}}$. Set parameter

$$\zeta_{\min} = \begin{cases} 1 & \text{if } \min_{s,t}(\bar{\mathbf{U}}_{s,t}) \geq -1 \\ \frac{1}{|\min_{s,t}(\bar{\mathbf{U}}_{s,t})|} & \text{otherwise} \end{cases} \tag{6.7}$$

2. Form an $m \times m$ matrix $\mathbf{T}^{NN}$ as

$$\mathbf{T}^{NN} = \frac{1}{n + \zeta_{\min}} \begin{bmatrix} 1 + \zeta_{\min} & 1 & \cdots & 1 \\ 1 & 1 + \zeta_{\min} & \cdots & 1 \\ \vdots & \vdots & \ddots & \vdots \\ 1 & 1 & \cdots & 1 + \zeta_{\min} \end{bmatrix} \tag{6.8}$$

3. The $(n \times m)$ matrix $\widetilde{\mathbf{U}} = \bar{\mathbf{U}}\mathbf{T}^{NN}$ is then SN and NN.

To see that $\bar{\mathbf{U}}\mathbf{T}^{NN}$ is SN, one notes that $\mathrm{sum}(\mathbf{T}^{NN}) = \mathbf{1}_{n\times1}$ and hence $\mathrm{sum}(\bar{\mathbf{U}}\mathbf{T}^{NN}) = \bar{\mathbf{U}}\mathbf{1}_{n\times1} = \mathrm{sum}(\bar{\mathbf{U}}) = \mathbf{1}_{n\times1}$ as $\bar{\mathbf{U}}$ is SN to begin with. To see that $\bar{\mathbf{U}}\mathbf{T}^{NN}$ is also NN, one notes that the tth column of $\bar{\mathbf{U}}\mathbf{T}^{NN}$ is given by the scalar $\frac{1}{n+\zeta_{\min}}$ multiplying a column vector which is the sum of $\mathbf{1}_{n\times1}$ and $\zeta_{\min}$ times the tth column of $\bar{\mathbf{U}}$, and the parameter $\zeta_{\min}$ ensures proper scaling to result in positive values for all entries.

The above shows that it is always possible to incorporate an additional NN condition to an SN matrix. Hence, a matrix can always be made SN first and then SN and NN. This further implies that a TP model can always be made to become SN and NN types, that is, a convex TP model.

6.2.3   Incorporating the NO condition

Similar to what we have achieved for SN and NN, we here desire invertible matrix $\mathbf{T}^{NO}$ of appropriate dimensions such that its product with the SN and NN matrix $\widetilde{\mathbf{U}}$,

$$\hat{\mathbf{U}} = \widetilde{\mathbf{U}}\mathbf{T}^{NO} \tag{6.9}$$

is NO. However, while it is always possible to tailor a matrix to satisfy SN and NN with the addition of at most one column, the same is not true for NO. Successful incorporation of the NO condition depends on the specific matrix at hand. Some matrices would require addition of many columns to become NO, the extreme case being until the matrix becomes full rank, which can then be transformed into an identity, and hence NO, matrix. From the computational complexity point of view, of course, the goal is to form a NO matrix while maintaining a small number of columns. Should this be not possible, we will have to settle for some milder versions of NO. The following gives a set of tight bounding procedures which yields a NO matrix if possible, and a CNO matrix otherwise.

Consider the $(n \times m)$ matrix $\widetilde{\mathbf{U}}$, which is SN and NN, from above. Noting that each of the n rows of $\widetilde{\mathbf{U}}$ corresponds to a point lying on an m-dimensional space and since $\widetilde{\mathbf{U}}$ is SN, these n points actually lie on a hyperplane of $(m-1)$ dimension. We present the following steps for tight bounding:

1. Project the n points in the m-dimensional space onto the $(m-1)$-dimensional hyperplane satisfying the SN condition. An efficient way to conduct the projection is to multiply $\widetilde{\mathbf{U}}$ on the right by the $m \times m$ matrix

$$
\begin{bmatrix}
1 & 0 & \cdots & 0 & 0 \\
1 & 1 & \cdots & 0 & 0 \\
\vdots & \vdots & \ddots & \vdots & \vdots \\
1 & 1 & \cdots & 1 & 0 \\
1 & 1 & \cdots & 1 & 1
\end{bmatrix}
\tag{6.10}
$$

The first column of the product will be 1s, as $\widetilde{\mathbf{U}}$ is SN. The remaining $(m-1)$ columns can be viewed as projected coordinates of the n points onto an $(m-1)$-dimensional plane.

2. Obtain the convex hull of the n points on the $(m-1)$-dimensional hyperplane. Algorithms to treat convex hull problems in a general dimensional space are discussed, for example, [O'R98] and [Ede87].

3. Check the convex hull. If the convex hull has exactly m vertices, successful incorporation of the NO condition is possible. In this case $\mathbf{T}^{NO}$ can be obtained as inverse of the matrix containing the m rows of $\widetilde{\mathbf{U}}$ associated with the convex hull. This is the NO case. If the convex hull has more than m vertices, however, determination of $\mathbf{T}^{NO}$ to strictly satisfy the NO condition is not possible. In this case, we have to search for a relaxed bounding with m vertices, not all of which came from the n points of $\widetilde{\mathbf{U}}$. The corresponding $\mathbf{T}^{NO}$ is then determined according to these m vertices. This is the CNO case.

Carrying out the procedures for $\widetilde{\mathbf{U}}$, we have $\hat{\mathbf{U}} = \widetilde{\mathbf{U}}\mathbf{T}^{NO}$, a NO, or CNO, matrix. Note that if $\hat{\mathbf{U}}$ is NO, then the choice of

$$
\hat{\mathbf{U}}' = \hat{\mathbf{U}}\mathbf{X}_a
\tag{6.11}
$$

is also NO for any given invertible and NO matrix $\mathbf{X}_a$.

It is worth mentioning that for the special case of $m = 2$ and $m = n$, the tightest bound always exists. When $m = 2$, the n points above are mapped onto a straight line. The convex hull is hence determined by the two extreme points on the line. When $m = n$, the n points themselves constitute the vertices of the convex hull on the $(n-1)$-dimensional hyperplane, or, as mentioned before, the matrix becomes full rank and can always be transformed into an identity matrix, which is NO.

It is important to point out that the convex hull in the NO procedures above is different from that defined by the convexity of the TP model discussed in Section 6.1.

6.2.4 Incorporating the RNO condition

As can be seen above, the NO condition is difficult to be fully incorporated in general. We may have to settle for a milder version, CNO, which requires only some of the columns to have the maximum value of 1. In this section, we consider another milder version of NO, which is the RNO condition. RNO requires that all the columns in the matrix to acquire the same maximum value c, even though c may be smaller than 1. When $c = 1$, the RNO condition becomes the NO condition. The price to pay for RNO is, again, an increase in the number of columns.

Consider the $(n \times m)$ matrix $\widetilde{\mathbf{U}}$ after the incorporation of SN and NN conditions. We desire to have matrix $\mathbf{P}$ of appropriate dimensions such that

$$\widehat{\mathbf{U}}\mathbf{P} = \widetilde{\mathbf{U}}$$

with matrix $\widehat{\mathbf{U}}$ being RNO. Notice that the formulation here is different from (6.9) for the incorporation of the NO condition. Here, $\mathbf{P}$ is actually not a square matrix and hence its inverse does not exist. There is no unique way to find matrix $\mathbf{P}$ to make $\widehat{\mathbf{U}}$ RNO. One way to do so is the following. Take $\widetilde{\mathbf{U}}$ as an example and denote its m columns as

$$\widetilde{\mathbf{U}} = \left[\widetilde{\mathbf{U}}_1 \,|\, \widetilde{\mathbf{U}}_2 \,|\, \ldots \,|\, \widetilde{\mathbf{U}}_m\right].$$

Then, for each $\widetilde{\mathbf{U}}_i$, we add a complementary column and a transformation matrix $\mathbf{T}_i$ to form a two-column component $\widehat{\mathbf{U}}_i$ that satisfies the SN, NN, and NO conditions:

$$\widehat{\mathbf{U}}_i = \left[\widetilde{\mathbf{U}}_i \quad (\mathbf{1}_{n \times 1} - \widetilde{\mathbf{U}}_i)\right] \mathbf{T}_i.$$

The transformation matrix $\mathbf{T}_i$ is given by

$$\mathbf{T}_i = \frac{1}{\overline{f}_i - \underline{f}_i} \begin{bmatrix} 1 - \underline{f}_i & \overline{f}_i - 1 \\ -\underline{f}_i & \overline{f}_i \end{bmatrix},$$

where $\underline{f}_i$ and $\overline{f}_i$ are the minimum and maximum values, respectively, of the elements of column $\widetilde{\mathbf{U}}_i$. As a result, we have $\widetilde{\mathbf{U}} = \widehat{\mathbf{U}}\mathbf{P}$, where

$$\widehat{\mathbf{U}} = \frac{1}{m}\left[\widehat{\mathbf{U}}_1 \,|\, \widehat{\mathbf{U}}_2 \,|\, \ldots \,|\, \widehat{\mathbf{U}}_m\right], \tag{6.12}$$

$$\mathbf{P} = m \begin{bmatrix} \mathbf{T}_1^{-1} & \mathbf{0}_{2\times2} & \cdots & \mathbf{0}_{2\times2} \\ \mathbf{0}_{2\times2} & \mathbf{T}_2^{-1} & \cdots & \mathbf{0}_{2\times2} \\ \vdots & \vdots & \ddots & \vdots \\ \mathbf{0}_{2\times2} & \mathbf{0}_{2\times2} & \cdots & \mathbf{T}_m^{-1} \end{bmatrix} \begin{bmatrix} 1 & 0 & \cdots & \cdots & 0 \\ 0 & 0 & \cdots & \cdots & 0 \\ 0 & 1 & \cdots & \cdots & 0 \\ 0 & 0 & \cdots & \cdots & 0 \\ \vdots & \vdots & \ddots & & \vdots \\ \vdots & \vdots & & \ddots & \vdots \\ 0 & 0 & \cdots & \cdots & 1 \\ 0 & 0 & \cdots & \cdots & 0 \end{bmatrix}. \tag{6.13}$$

Notice that each $\widehat{\mathbf{U}}_i$ is NO for $i = 1, 2, \ldots, m$, and the overall matrix $\widehat{\mathbf{U}}$ is SN, NN, and RNO with a constant $c = \frac{1}{m}$. As constructed, $\mathbf{P}$ is a $2m \times m$ matrix, and $\widehat{\mathbf{U}}$ is a $n \times 2m$ matrix. Hence, effectively, we have increased the number of columns by a factor of two for variable a.

As a final point, here we have initiated the RNO procedures with $\widetilde{\mathbf{U}}$ and $\widetilde{\mathbf{R}}$ after the SN and NN incorporation. If we so desire, the procedures can also start with $\overline{\mathbf{U}}$ and $\overline{\mathbf{R}}$ after SN, NN, and CNO incorporation.

6.3 Alternate method for INO and RNO conditions

The previous sections proposed algorithms to yield SN, NN, NO, CNO, and RNO type matrices. This section presents algorithms for INO and RNO transformations based on the work of [Bar06a]. The method introduced in the previous section doubles the columns of the resulting RNO type matrices. By contrast, the method to be introduced in this section does not increase the columns of the resulting matrix. As the new method is iterative, we need to redefine some of the notations. Let the $(n \times n)$ matrix obtained from HOSVD of (6.3), denoted as $\mathbf{U}$ from Section 6.2 onward, be labeled as $\mathbf{U}^{(0)}$ here. Moreover, denote also the retained matrix $\mathbf{U}^{(r)}$ as $\mathbf{U}^{(1)}$, and the matrix incorporating SN and NN conditions, $\widetilde{\mathbf{U}}$, as $\mathbf{U}^{(2)}$. The size of the matrix $\mathbf{U}^{(1)}$ is $n \times n^{(r)}$ and that of $\mathbf{U}^{(2)}$ is $n \times r$, where $r = n^{(r)}$ or $r = (n^{(r)} + 1)$, as the case may be. The previous sections have presented the construction of a transformation between $\mathbf{U}^{(1)}$ and $\mathbf{U}^{(2)}$, denoted as $\Theta^{(1)}$, such that

$$\mathbf{U}^{(1)} = \mathbf{U}^{(2)}\Theta^{(1)}. \tag{6.14}$$

Here, we present a new algorithm to construct an $(r \times r)$ transformation matrix $\Theta^{(2)}$ such that

$$\mathbf{U}^{(2)} = \check{\mathbf{U}}\Theta^{(2)}, \tag{6.15}$$

where $\check{\mathbf{U}}$ is our desired outcome, which is INO and RNO and of the same size as $\mathbf{U}^{(2)}$.

To make our steps more illustrative, we present a geometrical interpretation of Equation (6.15), in line with that in Section 6.2.3. Consider the rows of $\mathbf{U}^{(2)}$ and $\Theta^{(2)}$ as coordinates of points in a r-dimensional space. The special properties of the matrices have the following geometrical meaning:

- Since $\mathbf{U}^{(2)}$ and $\mathbf{U}$ are SN, $\Theta^{(2)}$ is also SN in (6.15). The fact that $\mathbf{U}^{(2)}$ and $\Theta^{(2)}$ are SN means that the points determined by their rows lie in an $(r - 1)$-dimensional hyperplane of the r-dimensional space. We can project all of the points to the $(r-1)$-dimensional space by multiplying $\mathbf{U}^{(2)}$ and $\Theta^{(2)}$ on the right with the matrix in (6.10), or by erasing the last column of $\mathbf{U}^{(2)}$ and $\Theta^{(2)}$ using the SN condition, to result in $\overline{\mathbf{U}}^{(2)}$ and $\overline{\Theta}^{(2)}$. The $(r-1)$-dimensional points determined by $\overline{\mathbf{U}}^{(2)}$ and $\overline{\Theta}^{(2)}$ are denoted as U_i and T_i, respectively.

- The r points of T_is form a simplex in the $(r-1)$-dimensional space.

- The relation (6.15) thus implies that the $(r-1)$-dimensional points U_i are a linear combination of the $(r-1)$-dimensional points T_i. Moreover, the coefficients of the linear combination are the elements of $\check{\mathbf{U}}$, which are all nonnegative due to the INO property of $\check{\mathbf{U}}$.

- All U_i points are hence inside the convex hull of the points T_i, which is the $(r-1)$-dimensional simplex. As there is a zero element in all the columns of $\check{\mathbf{U}}$, this means that each of the faces of this simplex contains at least one of the U_i points.

- The RNO property of $\check{\mathbf{U}}$ further implies that the maximum value of each column is the same value of $c \leq 1$. This means that there are r of the U_i points located with similar "proximity" to the r vertices of the simplex formed by T_i.

As noted before, for the case $r = 2$ and $r = n$, conversion of $\mathbf{U}^{(2)}$ to NO, hence, INO and RNO, is always possible. We are hence interested in the construction of $\Theta^{(2)}$ only for the case $r \geq 3$.

6.3.1   The partial algorithm

Let a be a nonnegative scalar and let $U_{i,j}$ denote the (i, j)th element of $\overline{\mathbf{U}}^{(2)}$. We regard the following transformations yielding the $U_{i,j}^D(a)$ elements (see Figure 6.2):

$$A: \quad U_{i,j}^A(a) = U_{i,j} + \frac{a-1}{r-1}\left(\sum_{k=1}^{r-1} U_{i,k} - 1\right)$$

$$B: \quad U_{i,j}^B(a) = U_{i,j}^A(a) - \min_l(U_{l,j}^A(a))$$

$$(6.16)$$

$$C: \quad U_{i,j}^C(a) = \frac{U_{i,j}^B(a)}{\max_l(U_{l,j}^B(a))}$$

$$D: \quad U_{i,j}^D(a) = \frac{U_{i,j}^C(a)}{\max_l(\sum_{k=1}^{r-1} U_{l,k}^C(a))}$$

We add an rth column to the $(n \times (r-1))$ matrix of the $U_{i,j}^D(a)$ values to make it SN. The resulting matrix is INO; the first $(r-1)$ columns are INO because of (6.16B) and the last one is INO because of (6.16D). Also, it is almost RNO, because the maxima of the $1, 2, \ldots, (r-1)$th columns are equal (which follows from the transformation

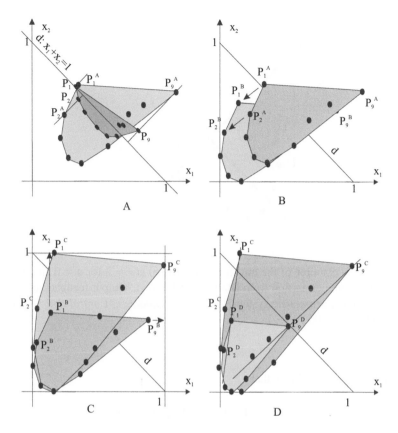

Figure 6.2: The four steps of the transformation (6.16) illustrated by an example with $r = 3$ and $n = 9$. (Replicated with permission from [Bar06a]–P. Baranyi,"Output feedback control of two-dimensional aeroelastic system," *Journal of Guidance, Control and Dynamics*, 29(3):762–767, 2006.) A: Perpendicular stretching by a from the line d ($a = 4$ in the figure), which would mean projection to d if $a = 0$. B: Shifting to the axis. C: Perpendicular stretching from the two coordinate axes to fill the unit square. D: Enlarging from the origin to hit line d.

(6.16C)). The maximum of the rth column may be different, depending on a. The RNO condition is satisfied if the maxima of the rth and the first columns are equal, that is, if their difference $f(a)$ is zero:

$$f(a) = 1 - \min_{l}\left(\sum_{k=1}^{r-1} U_{l,k}^D(a)\right) - \max_{m}\left(U_{m,l}^D(a)\right) = 0. \qquad (6.17)$$

Notice that

$$\lim_{a\to\infty} f(a) = 1 - 0 - \frac{1}{r-1} > 0 \qquad (6.18)$$

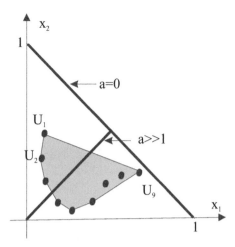

Figure 6.3: An example of the transformation (6.16) at $r = 3$ for $a = 0$ and $a \gg 1$. (Replicated with permission from [Bar06a]–P. Baranyi, "Output feedback control of two-dimensional aeroelastic system," *Journal of Guidance, Control and Dynamics*, 29(3):762–767, 2006.)

and in the case of $r = 3$ also that

$$f(0) = 1 - 1 - \max_m \left(U^D_{m,l}(0) \right) < 0 \tag{6.19}$$

(see Figure 6.3). Thus, if $r = 3$, Equation (6.17) has a positive solution a because the function $f(a)$ is continuous. We can solve Equation (6.17) numerically and produce the matrix $\check{U}$ satisfying the INO and RNO conditions. If $r > 3$, we need a more complex algorithm as discussed below.

6.3.2   The complete algorithm

If $r > 3$, the only difference is that Equation (6.19) does not necessarily hold. At the same time, if the $n \times (r - 1)$ matrix of the values $U^A_{i,j}(0)$ happens to be RNO and INO (it is always SN), then the result of the transformations (6.16B) to (6.16D) is identity and (6.19) is again true. Thus, we only need to apply an initial transformation on $\overline{U}^{(2)}$, which assures that the above matrix is RNO and INO. Finding this transformation is analogous to the original problem, but this matrix has only $(r - 1)$ columns instead of r. With such steps, the problem can be traced back to the case where $r = 3$, which has already been solved in the above Section 6.3.1. Now we describe the complete algorithm in detail.

Apply the following notations:

$$\overline{U}^{(2,r)} = \overline{U}^{(2)}, \tag{6.20}$$
$$U^{(2,r)} = U^{(2)} \tag{6.21}$$

and repeat the following simple recursion $r - 3$ times with $k = r - 1, r - 2, \ldots, 3$, respectively:

- Project the points determined by the matrix $\overline{U}^{(2,k+1)}$ to the hyperplane with the equation

$$\sum_{j=1}^{k} x_j = 1 \tag{6.22}$$

by the following transformation:

$$U_{i,j}^{(2,k)} = \overline{U}_{i,j}^{(2,k+1)} - \frac{1}{k}(\sum_{k=1}^{k} \overline{U}_{i,k}^{(2,k+1)} - 1) \tag{6.23}$$

- The matrix $\mathbf{U}^{(2,k)}$ is SN. Delete its last column, yielding the $n \times (k - 1)$ matrix $\overline{U}^{(2,k)}$.

The result is the $n \times 3$ size SN matrix $\mathbf{U}^{(2,3)}$. This can be transformed to become INO and RNO as described in the partial algorithm above. The result is denoted by $\mathbf{U}^{(4,3)}$. Then repeat the following recursive steps ($r - 3$) times with $k = 4, 5, \ldots, r$, respectively:

- Create the following matrix:

$$\overline{U}^{(3,k)} = \mathbf{U}^{(4,k-1)} - \mathbf{U}^{(2,k-1)} + \overline{U}^{(2,k)} \tag{6.24}$$

- Add a kth column to $\overline{U}^{(3,k)}$ to make it SN. (The result is $\mathbf{U}^{(3,k)}$.)

- Apply the transformations described in Section 6.3.1 on $\mathbf{U}^{(3,k)}$ to create the INO and RNO matrix $\mathbf{U}^{(4,k)}$. The algorithm gives a result in all steps because if $a = 0$, the transformation (6.16A) applied on $\overline{U}^{(3,k)}$ results in the INO and RNO matrix $\mathbf{U}^{(4,k-1)}$.

The matrix $\check{U} = \mathbf{U}^{(4,r)}$ is INO and RNO and it is constructed from $\mathbf{U}^{(2)}$ via an invertible inhomogenous linear transformation.

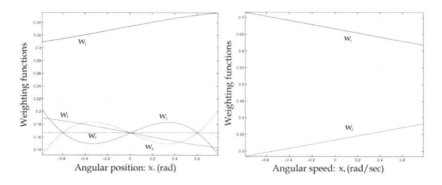

Figure 6.4: Weighting functions of the convex TP model 1 for x_3 and x_4.

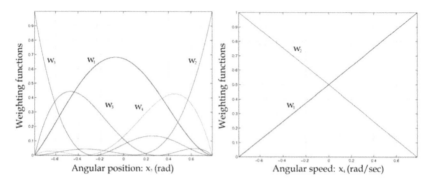

Figure 6.5: Weighting functions of the CNO type convex TP model 2 for x_3 and x_4.

6.4   The TORA example

Let us continue with the example of the TORA system studied in Section 4.3. The goal here is to transform the HOSVD-based canonical form of the TORA system to incorporate different convexity conditions. We pick the case when four singular values are kept on the dimension of $p_1 = x_3$ (see Figure 5.8) for our study.

- **TP model 1:** The first case study is to obtain a convex TP model of the TORA system by incorporating the SN and NN conditions. The resulting weighting functions are as depicted in Figure 6.4. Note that the number of weighting functions in the x_3 dimension is now five, which is one more than those in Figure 5.8. This is due to the additional column needed to enable the SN condition. Note that, however, the approximation error of the present TP model remains the same as that of Figure 5.8.

- **TP model 2:** The second case study is to define a tight convex hull of the TP

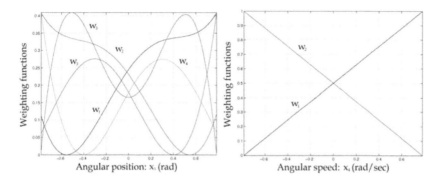

Figure 6.6: Weighting functions of INO and RNO type convex TP model 3 for x_3 and x_4.

model of the TORA system via generating CNO type weighting functions by proper transformation. The result is depicted in Figure 6.5. Note that not all weighting functions can achieve the maximum value of 1 in the x_3 dimension as it is only CNO. On the other hand, NO is achieved for the x_4 dimension as there are only two weighting functions there.

- **TP model 3:** The third task is to apply the algorithm in Section 6.3.2 to incorporate the INO and RNO conditions. The resulting weighting functions are depicted in Figure 6.6. Note that in this case all weighting functions of x_3 acquired the minimum values of 0 and the maximum value of roughly 0.42. Moreover, the number of weighting functions remains five after the incorporation.

Note that the above incorporation of convexity conditions may be derived analytically for some special cases but it would be extremely time consuming and tedious to achieve, if at all possible. Using the TP model transformation, however, the above manipulation can be achieved in a few minutes on a regular PC independently on the actual analytical representation of the given qLPV model. Moreover, should the model be modified, we can simply execute the computation again. This contrasts with the case of analytical derivations, which may require a completely new derivation given even slight changes to the given model. This illustrates the advantages of the proposed formulation.

Chapter 7

Introduction to the *TPtool* Toolbox

TPtool is a MATLAB Toolbox downloadable at "http://tptool.sztaki.hu" for implementing the algorithms of tensor product (TP) model transformation and the subsequent control system design. In this chapter, we will demonstrate the commands to generate the TP canonical form of a given quasi-linear parameter-varying (qLPV) model and incorporate the various types of convexities. The goal is to show how easily and readily applicable *TPtool* can be used to execute the TP model transformation and convex hull manipulation. More details of the commands are available through the command `help` in the *TPtool*. Algorithms dealing with the control system design will be demonstrated later via the application examples in the book.

7.1   Generating the TP canonical model

The steps of obtaining the TP canonical form are illustrated with the following simple qLPV system:

$$\dot{\mathbf{x}}(t) = \mathbf{S}(\mathbf{x}(t))\mathbf{x}(t) \tag{7.1}$$

where

$$\mathbf{S}(\mathbf{p}(t)) = \begin{pmatrix} 2\cos(4x_1(t)) & x_2(t) \\ 3 & x_2^3(t)\sin(x_1(t)) \end{pmatrix}. \tag{7.2}$$

Here, the parameter-varying vector is $\mathbf{p}(t) = \begin{pmatrix} x_1(t) \\ x_2(t) \end{pmatrix}$. The desired higher-order singular value decomposition (HOSVD)-based canonical form is a finite element TP type polytopic model where $\mathbf{S}(\mathbf{p}(t))$ above is converted into the following form:

$$\mathbf{S}(\mathbf{p}(t)) = \sum_{i=1}^{I} \sum_{j=1}^{J} w_{1,i}(x_1(t))w_{2,j}(x_2(t))\mathbf{S}_{i,j} = \mathcal{S} \underset{n=1}{\overset{2}{\boxtimes}} \mathbf{w}_n(x_n(t)) \qquad (7.3)$$

To obtain the corresponding weighting functions $\mathbf{w}_n$ and linear time-invariant (LTI) vertex systems $\mathcal{S}$ numerically one has to go through two steps in *TPtool*.

Step 1. The first step in *TPtool* is to select the parameter domain of interest Ω. This is an important step since the TP type polytopic model will be valid within this range. The domains of interest can be defined separately for each dimension of the vector $\mathbf{p}(t)$. One other parameter that has to be defined is M, which is the number of sampling points. *TPtool* uses equidistant sampling points, so if one wants to increase the accuracy of the polytopic model a larger value for M needs to be chosen. The number of sampling points can be also separately defined for each dimension of the vector $\mathbf{p}(t)$.

Let both $x_1(t)$ and $x_2(t)$ be sampled in the $[-0.5, 0.5]$ interval over 35 points:

```
domain = [-0.5 0.5; -0.5 0.5];
gridsize = [35 35];
```

Step 2. The goal of this step is to identify the TP structure of the given qLPV model and find the minimal number of LTI components via HOSVD. To execute Step 2, one has to define the qLPV model for *TPtool* first. The qLPV model should be given as a cell array of functions to the toolbox. One can use MATLAB's anonymous function handles to describe the parameter-varying system. The qLPV model above can be defined as follows:

```
 qlpvModel
= {...
    @(p)2*cos(4*p(1))        @(p)p(2);
    @(p)3                    @(p)p(2)^3*sin(p(1));
};
```

Note that `qlpvModel` can also incorporate data from direct measurement, fuzzy rule sets, neural networks, or the output of any complex algorithms.

One needs to point out that the parameter dependency in qLPV models is usually sparse, meaning that most elements of the corresponding system matrix depend only on a few parameters, or none at all. To sample the constant elements and elements that do not actually depend on the parameter being sampled increases the computational load tremendously. On the other hand, sampling of these elements is totally unnecessary. To overcome this inefficiency, in *TPtool* a dep matrix is used to describe which elements depend on which parameters. A zero element in the matrix dep indicates that the element is a constant and no sampling will be conducted. The issue of sampling the system elements according to their parameter dependence to enhance computational efficiency will be discussed in more detail in Chapter 9. The properly assigned dep matrix for the present qLPV system is given bellow:

```
% qlpvModel{i,j} depends on p(k) if dep(i,j,k)==1
dep = zeros([size(qlpvModel) 2]);
dep(1,1,:) = [1 0];
dep(1,2,:) = [0 1]
dep(2,1,:) = [0 0];
dep(2,2,:) = [1 1];
```

Sampling the system over the assigned grid and then numerically conducting HOSVD are executed by the following commands:

```
data = sampling_lpv(qlpvModel, dep, domain, gridsize);
[S U] = hosvd_lpv(data, dep, gridsize);
plothull(U);
```

Referring to (7.3), S here contains the numerically reconstructed values of the vertex systems $S_{i,j}$ and the U the numerical values of the discretized weighting functions $w_{1,i}(x_1(t))$ and $w_{2,j}(x_2(t))$ at the sample points along $x_1(t)$ and $x_2(t)$, respectively. The number of weighting functions is $I = J = 3$ for both parameters. Denote the LTI system estimates as $\hat{S}_{i,j}$, and the linearly interpolated weighting functions based on the discretized values in U as $\hat{w}_{1,i}(x_1(t))$ and $\hat{w}_{2,j}(x_2(t))$. The HOSVD-based canonical form of the qLPV model generated numerically by sampling over 35×35 points is given by

$$\mathbf{S}(x_1(t), x_2(t)) \approx \sum_{i=1}^{3} \sum_{j=1}^{3} \hat{w}_{1,i}(x_1(t))\hat{w}_{2,j}(x_2(t))\hat{\mathbf{S}}_{i,j} \tag{7.4}$$

An important question at this stage is the accuracy of the numerically constructed HOSVD-based canonical form (7.4) over the original qLPV model (7.1). We can numerically check, over 1000 random points, say, in the parameter domain Ω the differences between the two using the command:

```
[emax emean] = tperror(qlpvModel, S, U, domain, 1000);
```

```
disp('maxerror:');
disp(emax);
disp('mean error:');
disp(emean);
```

In this case the maximum error is 0.0034 and the mean error is 0.0013, which means that in the given parameter domain of $[-0.5, 0.5] \times [-0.5, 0.5]$:

$$\|\mathbf{S}(\mathbf{p}) - \hat{S} \underset{n=1}{\overset{2}{\boxtimes}} \hat{\mathbf{w}}_n(x_n)\|_2 \approx 0.001$$

Generally, there are two types of errors contributing to the difference between the qLPV model and its HOSVD-based canonical form. The first is the error of the vertex systems, and the second is due to the (usually linear) interpolation of the weighting functions. The first type of error can be made small by keeping a sufficient number of singular values in the HOSVD process. This is very important since the linear matrix inequality (LMI)-based design of the controllers later will be based on the vertex systems (see Chapter 11). The second type of error can be made small by increasing the number of sampling points along the dimensions. In the case that the original qLPV model embeds an inherent TP model, the HOSVD process will yield a finite number of nonzero singular values, even if the number of sampling points adopted for the dimensions are high. Hence, by adopting a very dense grid for sampling and keeping all the resultant nonzero singular values in the process, one can attain zero error in the first type and negligible error in the second type. In the limit, one can claim that the HOSVD numerically reproduces the exact inherent TP model embedded in a qLPV model. Should the original qLPV model not embed a TP model, the HOSVD process can produce a canonical form approximating it. In this case, one has to optimize between the number of nonzero singular values to keep in order to arrive at a canonical form of acceptable approximation to the original model.

7.2   Incorporating convexity conditions

This is the last step of TP model transformation but it is also very important. As we can observe, the HOSVD-generated weighting functions in Figure 7.1 do not satisfy any convexity condition. As such, they are not immediately useful for LMI-based control design, which requires convex polytopic representation. In this regard, *TPtool* provides various algorithms to generate different types of convex representations as described in Chapter 6. Let us consider the same system as in the previous example and generate various convexities for it.

The two important convexity types are the normality/close-to-normality (NO/CNO) and the IRNO (i.e., inverse normality [INO] and relaxed normality

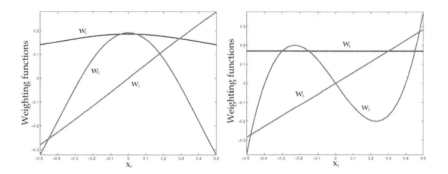

Figure 7.1: Canonical model weighting functions of a simple qLPV system.

[RNO]) types. As described in greater detail later, CNO, which yields a tighter convex hull, is generally best suited for controller design. On the other hand, IRNO is mostly suitable for observer design as it yields a large convex hull (see Chapter 12). One can also use a linear combination of CNO and IRNO convexity to achieve the best control outcome.

The desired convex TP models can be generated with the following function:

```
Ucno = genhull(U, 'cno');
Scno = coretensor(Ucno, data, dep);
plothull(Ucno);

Usnnn = genhull(U, 'snnn');
Ssnnn = coretensor(Usnnn, data, dep);
plothull(Usnnn);

Uirno = genhull(U, 'irno');
Sirno = coretensor(Uirno, data, dep);
plothull(Uirno);
```

Here, genhull is the function that generates the discretized values of the canonical model weighting function embedding the desired convexity. The input of genhull is U, which in this case is the numerical values of the discretized weighting functions $w_{1,i}(x_1)$ and $w_{2,j}(x_2)$ sampled along x_1 and x_2, respectively. The second parameter of genhull is the type of desired convexity to which to convert U.

On the other hand, the function coretensor computes the vertex systems corresponding to the desired convexity. The inputs are the discretized weighting functions of the canonical model embedding the desired convexity, the sampled data, and the dependency matrix.

It should be noted that the convex hull manipulation here is not done on the multidimensional S tensor but separately for each dimension via the associative n-mode matrices of the tensor as defined in Chapter 2.

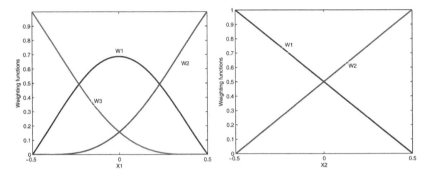

Figure 7.2: CNO weighting.

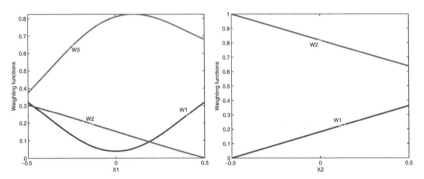

Figure 7.3: SNNN weighting.

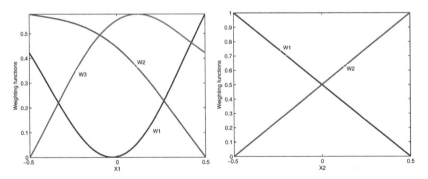

Figure 7.4: IRNO weighting.

The function plothull is to plot the weighting functions of the canonical model using the discretized values contained in the output of genhull.

The results of the plots can be seen in Figures 7.2–7.4. The convexity manipulation process can be shortened by a helper function:

```
[S U]=tptrans(lpv, dep, domain, gridsize, hull)
```

Chapter 8

Centralized Model Form

This chapter is mainly based on the work of [NBP08]. The primary objective is to introduce the centralization concept for the tensor product (TP) model and the canonical model of the quasi-linear parameter-varying (qLPV) systems. It will be shown using mathematical analysis and numerical examples that the proposed centralization form has important geometrical meanings which can be very useful in compact representation and enhanced computational performance.

8.1   The centralized model

Let the system matrix $\mathbf{S}(\mathbf{p})$ of a qLPV system be expressed in the TP model form of (2.11). We propose a centralized TP model form by subtracting a translation term $\mathbf{S}_0$ from the expression:

$$\mathbf{S}(\mathbf{p}) - \mathbf{S}_0 = \mathcal{S} \overset{N}{\underset{n=1}{\boxtimes}} \mathbf{w}_n(p_n) - \mathbf{S}_0. \tag{8.1}$$

With $\mathbf{S}_0$ being a TP model by itself, the right hand side of (8.1) is a linear combination of two TP models. As a linear combination of TP models can always be rewritten as a single TP model, $(\mathbf{S}(\mathbf{p}) - \mathbf{S}_0)$ is hence a TP model for arbitrarily given $\mathbf{S}_0$, that is,

$$\mathbf{S}(\mathbf{p}) - \mathbf{S}_0 = \mathcal{S}' \overset{N}{\underset{n=1}{\boxtimes}} \mathbf{w}'_n(p_n). \tag{8.2}$$

The class of systems describable by the centralized TP model is the same as that of the original TP model. The same can be proved for TP canonical form as well. If $\mathbf{S}(\mathbf{p})$ is originally expressed in a TP canonical form, then $(\mathbf{S}(\mathbf{p}) - \mathbf{S}_0)$ of (8.2) is also a TP canonical model.

We will show next that choosing a value S_0 different from zero can have appealing mathematical and numerical properties from the computational perspective.

8.1.1   Mathematical properties

Our goal is to describe $S(p)$ in a compact form. We will show that the centralized model allows more compact representation in a certain sense.

Definition 8.1 (Affine subspace). $A \subseteq V$ is an affine subspace of a vector space V over the field $\mathbb{R}$ if it is closed under affine combinations, that is, $\forall u, v \in A$ and $\forall t \in \mathbb{R}$, then $tu + (1 - t)v \in A$ as well.

Definition 8.2 (Affine closure). The affine closure of a set of vectors X is the affine subspace which contains all affine combinations of vectors in X. Let us denote this set with $\mathrm{aff}(X)$.

Note that if A is an affine subspace in V then $A - v = U$ is a subspace of V for $\forall v \in A$. So if $0 \in A$ then $A = U$, thus $\dim(A) = \dim(U)$ and if $0 \notin A$ then $-v \notin A - v$, thus $\dim(A) = \dim(U) + 1$. This means that translating a subset X of a vector space may result in a one-dimension reduction.

With the properties of affine closure and affine subspace, Theorem 8.1 can then be shown.

Theorem 8.1. $\forall X \subset V : \dim(X - v) = \dim(X) - 1$ *if and only if* $v \in \mathrm{aff}(X)$ *and* $0 \notin \mathrm{aff}(X)$.

The above turns out to be an important fact with respect to the TP model and the canonical model representation. Using the set $X = \{S(p)|p \in \Omega\}$, Theorem 8.1 shows that the representation of $S(p) - S_0$ may be simpler than the representation of $S(p)$, and thus the centralized model is more compact than the original model. The I_n number of weighting functions $w_{n,i}(p_n)$, $i = 1, \ldots, I_n$ of the TP model (2.11) can be reduced by one for each parameter dimension n in certain situations. A necessary condition for the reduction is using $S_0 \in \mathrm{aff}(X)$ for centralization.

Another approach to achieve a compact centralized model is to reduce its norm in some sense, and thus increase numerical accuracy and stability. A simple idea is to use Steiner's theorem to find S_0 for which the variance of the system is minimized:

$$E(S(p)) = \underset{S_0}{\mathrm{argmin}} \left(E((S(p) - S_0)^T (S(p) - S_0)) \right)$$

Thus, using the expected value of the system reduces the numeric values of the system.

8.1.2 Control properties

The linearized model of a nonlinear system has an important role in control theory. For the centralized model (8.1), one can choose S_0 to be the linearized model at a given operating point of the **p** parameter. Specifically, if the **p** parameter is a set of the state variables **x**, then S_0 can be the linearized model at the equilibrium point. With this choice of S_0, the centralized model thus assumes important control theoretic meaning. The weighting functions $w'_n(p_n)$ and the core tensor S' of (8.2) can be interpreted as the geometric directions of system deviation when the parameter **p** changes from the equilibrium point.

8.1.3 Computational advantages

Numerical reconstruction of the centralized model has the same basic properties as the original TP model or canonical model, since $(S(p) - S_0)$ in itself is a TP model or canonical form, as the case may be. However, using the centralized model has the following additional advantages:

- Using the centroid as S_0 results in numbers of reduced values in the core tensor; thus the computational algorithm can be numerically more precise.

- The memory usage of the algorithm may be possibly reduced. Theorem 8.1 shows that the centralized model may have one less weighting function for each of its parameters. In this case, the storage size of the discretized weighting function is reduced by NM, where N is the number of parameters and M is the sampling grid density, and the core tensor storage is approximately reduced by NJ^{N-1} where $J = I_1 = \cdots = I_N$ is the number of weighting functions for each parameter. This storage reduction factor will be even more beneficial in the case of future linear matrix inequality (LMI)-based controller designs.

8.2 Illustrating examples

We will use the TP canonical model to illustrate the above concepts. Consider a simple system:

$$S(x) = \begin{pmatrix} x & 100 \\ 0 & 0 \end{pmatrix} \tag{8.3}$$

It is easy to see that this system can be expressed as

$$S(x) = 100 \begin{pmatrix} 0 & 1 \\ 0 & 0 \end{pmatrix} + x \begin{pmatrix} 1 & 0 \\ 0 & 0 \end{pmatrix}, \tag{8.4}$$

showing its dependence on a constant function $w_1(x) = 100$ and a linear function
$w_1(x) = x$ of parameter x. These two functions are linearly independent, and thus the
TP canonical form has two weighting functions as well (see Figure 8.1). The output
of the numerical reconstruction algorithm of the canonical form by *TPtool* is

```
singular values:
        1000
        5.8315
number of singular values to keep [all] = 2
mean error:
 3.6045e-015
```

where we used the sampling domain $\Omega = [-1, 1]$ and grid size $M = 100$. As can
be seen, the first weighting function $w_1(x)$ has a larger singular value than the second. This is obvious since to represent the original system when $x \in [-1, 1]$ with
minimal error, the constant term is more important. However, the finer structure of
the x-parameter dependency of the system is buried under the constant term, which
constitutes the linearized model at the equilibrium point $x = 0$.

In the case of the centralized model, we can thus choose $\mathbf{S}_0$ as the linearized
model:

$$\mathbf{S}_0 = \mathbf{S}(0) = \begin{pmatrix} 0 & 100 \\ 0 & 0 \end{pmatrix}$$

so as to obtain information about the system behavior around the $x = 0$ point. The resulting $(\mathbf{S}(x) - \mathbf{S}_0)$ system then becomes simpler, and model size reduction is possible
since

$$\mathbf{S}(x) - \mathbf{S}_0 = x \begin{pmatrix} 1 & 0 \\ 0 & 0 \end{pmatrix}$$

so only one weighting function is needed. The output of the numerical reconstruction
algorithm is

```
singular values:
        5.8315
number of singular values to keep [all] = 2
mean error:
 1.0358e-017
```

Here we obtained the desired information with fewer calculations and greater precision. The resulting weighting function can be seen in Figure 8.2.

Let us consider a more involved system:

$$\mathbf{S}(x, y) = \begin{pmatrix} xy^2 + 100 & x^2y + 100 \\ x^3y + 100 & 100 \end{pmatrix} \tag{8.5}$$

To represent the system one needs at least four weighting functions for the x parameter and three weighting functions for the y parameter, since $[1, x, x^2, x^3]$ functions and $[1, y, y^2]$ functions are linearly independent and are obviously needed to represent the system.

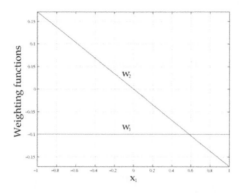

Figure 8.1: Weighting functions of the canonical model of the original model of system (8.3).

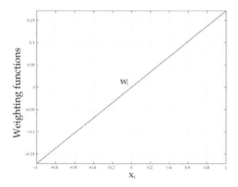

Figure 8.2: Weighting functions of the canonical model of the centralized model of system (8.3).

Figure 8.3 shows the four x-parameter weighting functions and three y-parameter weighting functions of the TP canonical model of the system. The output of the numerical reconstruction algorithm is:

```
-- First dimension
singular values:
        20000
       32.841
       17.735
```

```
        6.8257
number of singular values to keep [all] = 4

-- Second dimension
singular values:
        20000
        33.543
        17.735
number of singular values to keep [all] = 3

mean error:
  6.7787e-013
```

Using the centralized model form, the system becomes:

$$\mathbf{S}(x, y) - \begin{pmatrix} 100 & 100 \\ 100 & 100 \end{pmatrix} = \begin{pmatrix} xy^2 & x^2y \\ x^3y & 0 \end{pmatrix}$$

In this case, the size of the model is reduced. Figure 8.4 shows the three and two weighting functions for the x and y parameters, respectively, of the TP canonical model of the centralized model. The numerical stability is also improved, as the numerical reconstruction algorithm output shows much smaller mean error:

```
-- First dimension
singular values:
        34.26
        26.604
        7.0511
number of singular values to keep [all] = 3

-- Second dimension
singular values:
        34.978
        26.604
number of singular values to keep [all] = 2

mean error:
  2.9123e-016
```

As a summary, the centralized TP form proposed in this chapter has the advantage that it yields the same function class and basic properties as the original TP model or canonical model, as the case may be, while allowing for possible weighting function dimensionality reduction. This leads to relaxed demand on computational resources for the ensuing processing algorithms. Translating the original TP model

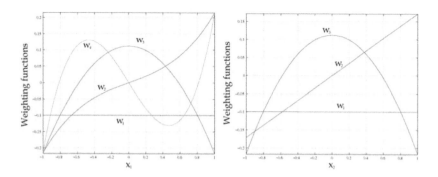

Figure 8.3: Weighting functions of the canonical model of the original model of system (8.5).

or canonical model to its centroid location also results in smaller numerical values, thus enhancing the numerical stability and precision. Furthermore, the quantities S_0 and the core tensor also provide certain control theoretical meanings and useful geometrical interpretation.

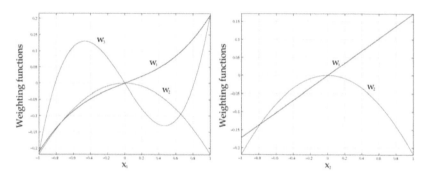

Figure 8.4: Weighting functions of the canonical model of the centralized model of system (8.5).

Chapter 9

Computational Relaxed TP Model Transformation

Any method that operates on multidimensional data suffers from the curse of dimensionality, that is, complexity increases exponentially but the algorithms are not scalable. Such a problem arises in the tensor product (TP) model transformation as well when the model depends on many parameters. The size of the sampled model increases exponentially with respect to the number of parameters, and the higher-order singular value decomposition (HOSVD) cannot cope with the huge amount of data.

A general solution to this problem is, of course, to enhance the available computational resources. In many cases, however, complexity reduction offers another route. Many control problems—and many other real world problems as well—do have special properties that enable taking such an approach. In this regard, an important property of the HOSVD is that it features a compact representation for execution, so there is some hope for finding an algorithm which can produce this compact HOSVD representation directly using a minimal amount of data at the discretization step. To achieve this, we should exploit the redundancy in the model, namely, the fact that elements in a system matrix may not all depend on every parameter, and thus it is not necessary to sample all elements in the whole parameter space. This idea can effectively reduce the size of the sampled system, since most quasi-linear parameter-varying (qLPV) models contain many constant elements and rarely contain elements that depend on every parameter.

This chapter introduces a modified TP model transformation algorithm that is based on the works of [NPBH09] which utilizes the complexity reduction outlined above. The main idea behind this modified, computationally relaxed algorithm is to partition the problem and handle the parameter-varying system matrix element by element. Sampling and executing HOSVD on the individual system element basis can significantly reduce the time and space complexity. The challenge of this method is the reconstruction of the original HOSVD of the whole sampled system from the element-wise data.

The following is an informal overview of the main ideas behind the modified method:

- Singular value decomposition (SVD) can be calculated by separately applying to parts of a matrix as shown in Figure 9.1. This means that SVD can be conducted in a distributed manner without the need to load all the data into memory.

- If a matrix has a simple known structure (e.g., it contains many equal elements), then its SVD can be conducted based on some short and efficient description of the matrix instead of its full numeric data set. Another example would be a matrix of rank 1 that can be expressed as a multiplication of two vectors, and it is straightforward to calculate the SVD from the vectors.

- Combining the two ideas above, certain simple structured parts of a large matrix can be calculated separately and efficiently, and the results can later be joined together to form the SVD.

- System matrices normally contain many constant elements or elements which depend on just a few parameters. Sampling these elements in the whole parameter space produces a data tensor of high information redundancy. This motivates the use of the n-mode matrix layouts of the data tensor in the modified method for a simpler and more efficient structure.

- The n-mode matrices mentioned can be obtained as follows. For each element in the system matrix, one conducts sampling only for the parameters on which the element is depending. This greatly reduces the amount of data being generated.

- Using the above, it is hence sufficient to sample one element of the system matrix at a time for only a few parameters, and then run the SVD algorithm on this relatively small amount of data before putting the pieces together properly.

Applications of these ideas are given in the following sections. Specifically, Section 9.1 presents the mathematical background mainly of the first two ideas above. Section 9.2 gives the formulation of the modified transformation and the corresponding TP model transformation algorithm. Section 9.3.2 further describes the relaxed computation requirement of the modified algorithm compared with the original transformation.

9.1 SVD-based column equivalence

The following theorems can be proved for complex matrices as well, but for simplification real matrices are assumed here. These theorems form the basis of our modified, computationally relaxed TP transformation here. They also provide a way of calculating the SVD of a large nonsquare matrix on a parallel and distributed process, or in situations when the whole matrix does not fit into the memory.

Definition 9.1 (SVD-based column equivalence). Let $A \in \mathbb{R}^{n \times m}$ and $B \in \mathbb{R}^{n \times k}$ be real matrices. They are *SVD-based column equivalent* ($A \sim B$) if there exists a compact singular value decomposition (SVD) $A = U_A S_A V_A^T$ and $B = U_B S_B V_B^T$ such that $U_A = U_B$ and $S_A = S_B$.

Compact SVD here means that the zero singular values are discarded together with the respective columns in U and V, and the singular values are in a descending order in S. Hereafter, SVD always refers to compact SVD.

Note that the SVD-based column equivalent matrices have the same number of rows, but they can have a different number of columns, since the equality of V^T matrices is not required. The definition is similar to range, image, or column space equivalence (see [GVL96]). However, SVD-based column equivalence poses a stronger condition since it requires the equivalence of the singular values as well.

Note also that the SVD factorization of a real matrix is *almost* unique (the sign and degenerate singular vectors may vary, but S is unique). Therefore, if $A \sim B$, then for *any* SVD of A there exists a corresponding SVD of B, where U and S are equal.

Intuitively, SVD-based column equivalence means that the column vectors of the two matrices have the same principal components—U is a basis set in the column space of both matrices and each basis vector has the same "significance" for the two matrices, that is, leaving out a basis vector from U gives the same quadratic error for both.

Lemma 5. SVD-based column equivalence is an equivalence relation.

Proof. All three requirements can be easily derived from the definition:

- reflexivity: $A \sim A$

- symmetry: if $A \sim B$ then $B \sim A$

- transitivity: if $A \sim B$ and $B \sim C$ then $A \sim C$ □

Lemma 6. Let A be a real matrix. Then, any reordering of its columns gives an SVD-based column equivalence matrix, that is, ($A \sim \pi(A)$ for any column permutation π).

Proof. We can get a valid SVD factorization of a reordered matrix by the same re-ordering of $\mathbf{V}^T$ and leave $\mathbf{U}$ and $\mathbf{S}$ unchanged.   □

Lemma 7. Let $\mathbf{A}$ be a real matrix with the SVD factorization $\mathbf{A} = \mathbf{USV}^T$. Then $\mathbf{A} \sim \mathbf{US}$.

Proof. A valid SVD factorization of $\mathbf{US}$ is $\mathbf{US} = \mathbf{USI}$ where $\mathbf{I}$ is an identity matrix of corresponding size.   □

Lemma 8. If $\mathbf{A} = \mathbf{xy}^T$ is a real matrix of rank 1, then $\mathbf{A} \sim \sqrt{\mathbf{y}^T\mathbf{y}}\mathbf{x}$.

$$\mathbf{A} \sim \sqrt{\mathbf{y}^T\mathbf{y}}\mathbf{x}.$$

Proof. $\mathbf{A} = \mathbf{xy}^T = \mathbf{u}\sqrt{\mathbf{x}^T\mathbf{x}}\sqrt{\mathbf{y}^T\mathbf{y}}\mathbf{v}^T$, where $\mathbf{u} = \frac{\mathbf{x}}{\sqrt{\mathbf{x}^T\mathbf{x}}}$ and $\mathbf{v} = \frac{\mathbf{y}}{\sqrt{\mathbf{y}^T\mathbf{y}}}$ are the normalized $\mathbf{x}$ and $\mathbf{y}$ respectively, so this is a valid SVD factorization. According to Lemma 7 $\mathbf{A} \sim \mathbf{u}\sqrt{\mathbf{x}^T\mathbf{x}}\sqrt{\mathbf{y}^T\mathbf{y}} = \sqrt{\mathbf{y}^T\mathbf{y}}\mathbf{x}$.   □

Corollary 9.1. *If* $\mathbf{A} = \mathbf{a}\begin{bmatrix} 1 & 1 & \dots & 1 \end{bmatrix}$ *consists of M number of the same* $\mathbf{a}$ *column, then* $\mathbf{A} \sim \sqrt{M}\mathbf{a}$.

Lemma 9. Let $[\mathbf{A}|\mathbf{B}]$ be a real matrix, with $\mathbf{A}$ and $\mathbf{B}$ as its submatrices (a partition of the column vector set). If $\mathbf{B} \sim \mathbf{C}$, then $[\mathbf{A}|\mathbf{B}] \sim [\mathbf{A}|\mathbf{C}]$ holds too.

Proof. Let the SVD factorization of $\mathbf{B}$ be $\mathbf{B} = \mathbf{USV}^T$. First, we show that

$$[\mathbf{A}|\mathbf{B}] \sim [\mathbf{A}|\mathbf{US}].$$

Let the SVD of $[\mathbf{A}|\mathbf{US}]$ be $[\mathbf{A}|\mathbf{US}] = \mathbf{U}'\mathbf{S}'[\mathbf{Z}^T|\mathbf{W}^T]$ where $\mathbf{US} = \mathbf{U}'\mathbf{S}'\mathbf{W}^T$. Then

$$[\mathbf{A}|\mathbf{B}] = [\mathbf{A}|\mathbf{USV}^T] = \mathbf{U}'\mathbf{S}'[\mathbf{Z}^T|\mathbf{W}^T\mathbf{V}^T] = \mathbf{U}'\mathbf{S}'\mathbf{V}'^T$$

also holds. This would be a valid SVD factorization if the columns of $\mathbf{V}' = [\mathbf{Z}^T|\mathbf{W}^T\mathbf{V}^T]^T$ were orthonormal.

The columns of a matrix $\mathbf{Q}$ are orthonormal if $\mathbf{Q}^T\mathbf{Q} = \mathbf{I}$. The columns of $[\mathbf{Z}^T|\mathbf{W}^T]^T$ are orthonormal (since it is an SVD component) and thus

$$[\mathbf{Z}^T|\mathbf{W}^T][\mathbf{Z}^T|\mathbf{W}^T]^T = \mathbf{Z}^T\mathbf{Z} + \mathbf{W}^T\mathbf{W} = \mathbf{I}$$

Similarly, we get

$$\mathbf{V}'^T\mathbf{V}' = \mathbf{Z}^T\mathbf{Z} + \mathbf{W}^T\mathbf{V}^T\mathbf{V}\mathbf{W} = \mathbf{Z}^T\mathbf{Z} + \mathbf{W}^T\mathbf{W} = \mathbf{I}, \qquad (9.1)$$

where we used the orthonormality of $\mathbf{V}$. This shows the orthonormality of $\mathbf{V}'$.

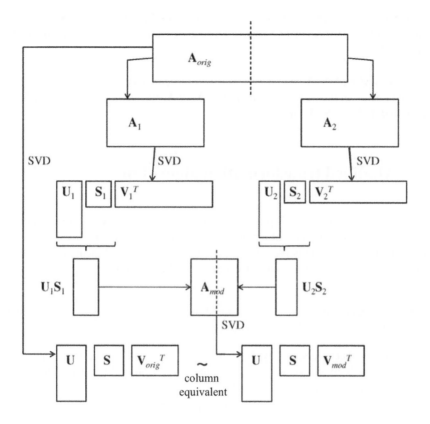

Figure 9.1: Parallel SVD decomposition: Illustration for Corollary 9.2. (Modified from [NPBH09]–S. Nagy, Z. Petres, P. Baranyi, and H. Hashimoto, "Computational relaxed TP model transformation: restricting the computation to subspaces of the dynamic model," *Asian Journal of Control*, 11(5):461–475, 2009.)

Thus, $[\mathbf{A}|\mathbf{B}] \sim [\mathbf{A}|\mathbf{US}]$. According to Lemma 7, the same can be proven for $[\mathbf{A}|\mathbf{C}]$ so

$$[\mathbf{A}|\mathbf{B}] \sim [\mathbf{A}|\mathbf{US}] \sim [\mathbf{A}|\mathbf{C}].$$

$\square$

Theorem 9.1. *For a set of matrices* $\mathbf{A}_1, \mathbf{A}_2, \ldots, \mathbf{A}_n$ *and* $\mathbf{B}_1, \mathbf{B}_2, \ldots, \mathbf{B}_n$, *if* $\mathbf{A}_i \sim \mathbf{B}_i$ *holds* $(i = 1, \ldots, n)$, *then* $[\mathbf{A}_1|\ldots|\mathbf{A}_n] \sim [\mathbf{B}_1|\ldots|\mathbf{B}_n]$.

Proof. It follows from Lemma 6 and Lemma 9. $\square$

Corollary 9.2. *Let* $\mathbf{A} \in \mathbb{R}^{n \times k}$ *be a large real matrix such that* $n \ll k$. *The SVD of* $\mathbf{A}$ *can be calculated by dividing it up into two (or more) submatrices* $\mathbf{A} = [\mathbf{A}_1|\mathbf{A}_2]$ *and then calculating SVD for each submatrix* $\mathbf{A}_i = \mathbf{U}_i\mathbf{S}_i\mathbf{V}_i^T$. *We can then obtain the*

U *and* **S** *matrices of the SVD of* **A** *by calculating the SVD of* $[\mathbf{U}_1\mathbf{S}_1|\mathbf{U}_2\mathbf{S}_2]$, *which is at most a* $n \times 2n$ *matrix (much smaller than the original). This method is illustrated in Figure 9.1.*

The row size of the **V** *matrix is k so it is a large matrix. If we want to calculate* **V** *we can do so in parts. For any submatrix* $\mathbf{A}_i$ *of* **A**, *the corresponding* $\mathbf{V}_i^T$ *of* $\mathbf{V}^T$ *is calculated as* $\mathbf{V}_i^T = \mathbf{S}^{-1}\mathbf{U}^T\mathbf{A}_i$.

9.2   Modified transformation algorithm

Assume that a qLPV matrix $\mathbf{S}(\mathbf{p})$ with the size $O \times I$ is given, where the parameter vector $\mathbf{p} \in \Omega$ has N number of elements. In the following, the modified TP model transformation algorithm is presented in contrast to the original method. Before describing the algorithm, let us introduce some useful definitions.

- Ω is, again, a hyper-rectangle in the N-dimensional parameter space as given by $[a_1, b_1] \times [a_2, b_2] \times \cdots \times [a_N, b_N]$.

- The grid net is defined by M_n number of equidistantly located points g_{n,m_n} along each dimension ($n = 1, \ldots, N$, $m_n = 1, \ldots, M_n$) of the parameter space. Therefore, a grid point is defined as $\mathbf{g}_{m_1,m_2,\ldots,m_N} = \begin{pmatrix} g_{1,m_1} & \cdots & g_{N,m_N} \end{pmatrix}$. Denote the total number of grid points as $G = \prod_{n=1}^{N} M_n$.

- Denote an element of the system matrix $\mathbf{S}(\mathbf{p})$ by $S_{o,i}(\mathbf{p})$ and its discretized tensor over the whole Ω by $\mathcal{S}_{o,i}^d \in \mathbb{R}^{M_1 \times M_2 \times \cdots \times M_N \times O \times I}$, that is,

$$(\mathcal{S}_{o,i}^d)_{m_1,m_2,\ldots,m_N} = S_{o,i}(\mathbf{g}_{m_1,m_2,\ldots,m_N}) \in \mathbb{R}$$

- Denote the dependent dimensions of the $S_{o,i}$ element by $J_{o,i} \subseteq \{1,\ldots,N\}$, that is, if $S_{o,i}(\mathbf{p})$ depends on the p_j parameter, then $j \in J_{o,i}$.

- Denote by $\mathcal{T}_{o,i}$ the discretized tensor of $S_{o,i}(\mathbf{p})$ sampled only in the dependent dimensions $J_{o,i}$. Thus, $\mathcal{T}_{o,i} \in \mathbb{R}^{M_{j_1} \times M_{j_2} \times \ldots M_{j_k}}$, with $\{j_1, j_2, \ldots, j_k\} = J_{o,i}$ and

$$\mathcal{S}_{o,i}^d = \mathcal{T}_{o,i} \underset{n \notin J_{o,i}}{\boxtimes} \mathbf{1}_{M_n \times 1},$$

where $\mathbf{1}_{M_n \times 1}$ is the $M_n \times 1$ vector containing ones. This means that $\mathcal{S}_{o,i}^d$ contains copies of $\mathcal{T}_{o,i}$ along the nondependent (and thus nondiscretized) dimensions.

- We use the following notation for the n-mode matrices of the subtensors:

$$\mathbf{H}_n^{o,i} = (\mathcal{S}_{o,i}^d)_{(n)}$$

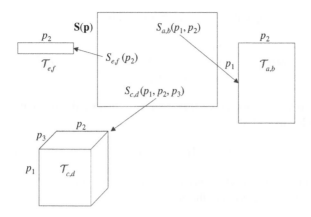

Figure 9.2: Sampling a system matrix element-wise. (Modified from [NPBH09]–
S. Nagy, Z. Petres, P. Baranyi, and H. Hashimoto, "Computational relaxed TP model
transformation: restricting the computation to subspaces of the dynamic model,"
Asian Journal of Control, 11(5):461–475, 2009.)

$$\mathbf{K}_n^{o,i} = (\mathcal{T}_{o,i})_{(n)}$$

Note that if n is not a sampled dimension (i.e., $n \notin J_{o,i}$), then the n-mode matrix
$\mathbf{K}_n^{o,i}$ is a row vector containing every element of $\mathcal{T}_{o,i}$, since the size of $\mathcal{T}_{o,i}$ in
that dimension is 1.

The modified TP model transformation can now be conducted in the following
three steps corresponding to those of the original version in Section 3.2.

Step 1: Discretization

The discretization step differs from the original in the modified TP model transformation. In the original case, the whole system matrix is discretized over all grid
points as:

$$\mathbf{S}_{m_1,\dots,m_N}^d = \mathbf{S}(\mathbf{g}_{m_1,m_2,\dots,m_N}),$$

the $O \times I$-sized $\mathbf{S}_{m_1,\dots,m_N}^d$ matrices are stored into tensor $\mathcal{S}^d \in \mathbb{R}^{M_1 \times \cdots \times M_N \times O \times I}$.

However, in the case that not all elements of the qLPV system matrix $\mathbf{S}(\mathbf{p})$ depend
on all elements of the parameter vector, we can discretize inidvidual element $S_{o,i}(\mathbf{p})$
($o = 1, \dots, O, i = 1, \dots, I$) of $\mathbf{S}(\mathbf{p})$ separately and only for the respective elements of
$\mathbf{p}$ it depends upon as contained in the dependent dimensions $J_{o,i}$. This is illustrated in
Figure 9.2. With this discretization, we obtain the subtensor $\mathcal{T}_{o,i}$, the order of which
is equal to $|J_{o,i}|$.

Consider for example the system matrix given by:

$$\mathbf{S}(\mathbf{p}) = \begin{pmatrix} S_{1,1} & S_{1,2}(p_1, p_2) & S_{1,3} \\ S_{2,1} & S_{2,2} & S_{2,3}(p_2, p_3) \\ S_{3,1}(p_1, p_3) & S_{3,2}(p_2) & S_{3,3} \end{pmatrix}, \tag{9.2}$$

where vector $\mathbf{p}$ has the elements p_1, p_2, and p_3. When we discretize the elements we obtain the following subtensors:

$$\mathcal{T}_{1,2} \in \mathbb{R}^{M_1 \times M_2} \quad \mathcal{T}_{2,3} \in \mathbb{R}^{M_2 \times M_3} \quad \mathcal{T}_{3,1} \in \mathbb{R}^{M_1 \times M_3} \quad \mathcal{T}_{3,2} \in \mathbb{R}^{M_2}$$

and the remaining constant elements are stored in corresponding subtensors $\mathcal{T}_{o,i} \in \mathbb{R}$. This example will be referred to throughout this section.

Step 2: Extraction of the TP structure via HOSVD

This step involves the generation of three components, namely, the weighting functions, the HOSVD core tensor, and the continuous weighting functions.

Determining the weighting functions   The original TP model transformation executes HOSVD on the first N dimension of the discretized tensor $\mathcal{S}^d \in \mathbb{R}^{M_1 \times \cdots \times M_N \times O \times I}$, discarding the zero singular values (or small values in case of trade-off) and corresponding singular vectors during the process. For each dimension, we first generate the n-mode matrix $(\mathcal{S}^d)_{(n)} = \mathbf{H}_n$ of the discretized tensor $\mathcal{S}^d$, and then conduct SVD on $\mathbf{H}_n$ to obtain the matrix $\mathbf{U}_n$.

In the modified method, the HOSVD is computed along the dimensions as in the case of the original TP model transformation. The difference is that we obtain the matrix $\mathbf{U}_n$ by conducting SVD of the matrix $\mathfrak{H}_n$, which is an *SVD-based column equivalent* (Definition 9.1) of $\mathbf{H}_n$ and normally much smaller in size. The modified version is thus more efficient in terms of computation and storage complexity.

Specifically, we generate $\mathfrak{H}_n \sim \mathbf{H}_n$ from the elements of the system matrix. This is feasible as a possible layout of the $\mathcal{S}^d$ tensor as

$$\mathbf{H}_n = [\mathbf{H}_n^{1,1} | \mathbf{H}_n^{1,2} | \dots | \mathbf{H}_n^{O,I}]$$

and according to Theorem 9.1, a possible *SVD-based column equivalent* matrix is

$$\mathfrak{H}_n = [\mathfrak{H}_n^{1,1} | \mathfrak{H}_n^{1,2} | \dots | \mathfrak{H}_n^{O,I}]$$

where $\mathfrak{H}_n^{o,i} \sim \mathbf{H}_n^{o,i}$. To calculate $\mathfrak{H}_n^{o,i}$ from the given $\mathcal{T}_{o,i}$ we should distinguish between two cases: laying out the subtensor $\mathcal{T}_{o,i}$ in a discretized (i.e., dependent, $n \in J_{o,i}$) dimension or a nondiscretized (i.e., independent, $n \notin J_{o,i}$) dimension. Here, we give the formulas for the two cases.

- The $n \in J_{o,i}$ case:

$$\mathbf{H}_n^{o,i} = [\mathbf{K}_n^{o,i} | \mathbf{K}_n^{o,i} | \dots | \mathbf{K}_n^{o,i}]$$

since $\mathcal{S}_{o,i}^d$ just contains copies of $\mathcal{T}_{o,i}$ in every nondiscretized dimension, the number of copies is the size of the nondiscretized dimension, that is, $\prod_{m \notin J_{o,i}} M_m$.

Reordering the n-mode matrix so that the same columns are next to each other shows that

$$\mathbf{H}_n^{o,i} \sim \sqrt{\prod_{m \notin J_{o,i}} M_m} \mathbf{K}_n^{o,i} = \mathfrak{H}_n^{o,i}, \tag{9.3}$$

where we used Corollary 9.1 and Lemma 6.

If the dimensionality of $\mathbf{K}_n^{o,i}$ is big (i.e., $|J_{o,i}| > 2$), this reduction may not be enough. One may need to use a *minimal SVD-based column equivalent* form by letting $\mathbf{K}_n^{o,i} = \mathbf{U}_K \mathbf{S}_K \mathbf{V}_K$ be an SVD factorization according to Lemma 7, and then

$$\mathbf{H}_n^{o,i} \sim \sqrt{\prod_{m \notin J_{o,i}} M_m} \mathbf{K}_n^{o,i} \sim \sqrt{\prod_{m \notin J_{o,i}} M_m} \mathbf{U}_K \mathbf{S}_K = \mathfrak{H}_n^{o,i}. \tag{9.4}$$

The transformation (9.4) does not increase the overall complexity and it reduces memory usage. Its numerical precision, however, is lower than simply using $\mathbf{K}_n^{o,i}$ with a constant multiplier as in (9.3).

- The $n \notin J_{o,i}$ case: In this case, the n-mode matrix is given by

$$\mathbf{H}_n^{o,i} = \mathbf{1}_{M_n \times 1} [\mathbf{K}_n^{o,i} | \mathbf{K}_n^{o,i} | \dots | \mathbf{K}_n^{o,i}]$$

where $\mathbf{1}_{M_n \times 1}$ is the $M_n \times 1$ constant column vector containing ones and $\mathbf{K}_n^{o,i}$ is a row vector.

The number of copies of $\mathbf{K}_n^{o,i}$ is the size of the nondiscretized dimensions minus one (the concerned dimension n itself), that is, $\prod_{m \notin \{J_{o,i}, n\}} M_m$. Using Corollary 9.1 and Lemma 6 again, we get

$$\mathbf{H}_n^{o,i} \sim \sqrt{\prod_{m \notin \{J_{o,i}, n\}} M_m} \mathbf{1}_{M_n \times 1} \left\| \mathcal{T}_{o,i} \right\| = \mathfrak{H}_n^{o,i} \tag{9.5}$$

The n-mode matrix of the $\mathcal{S}^d$ tensor is then

$$\mathbf{H}_n = [\mathbf{H}_n^{1,1} | \mathbf{H}_n^{1,2} | \dots | \mathbf{H}_n^{O,I}] \sim [\mathfrak{H}_n^{1,1} | \mathfrak{H}_n^{1,2} | \dots | \mathfrak{H}_n^{O,I}] = \mathfrak{H}_n$$

With these formulas, we can easily calculate a $\mathfrak{H}_n^{o,i} \sim \mathbf{H}_n^{o,i}$ from $\mathcal{T}_{o,i}$. Then, according to Theorem 9.1, $\mathfrak{H}_n \sim \mathbf{H}_n$ so by the definition of SVD-based column equivalence, $\mathbf{U}_n$ can be obtained from the SVD of $\mathfrak{H}^{(n)}$.

Note that special transformations on the TP model are still needed in order to fulfill certain requirements for controller design. These are the transformations, for example, to generate convex TP models satisfying the conditions of sum normalization (SN) and nonnegativeness (NN) on $\mathbf{U}_n$, which have been discussed in Chapter 6.

For illustration of the above algorithm, consider the example (9.2) again. To generate the matrix $\mathbf{U}_1$, we need to determine the SVD-based column equivalent matrix of the first-mode matrix $\mathbf{H}_1$, $\mathfrak{H}_1$, which is formed by the submatrices $\mathfrak{H}_1^{o,i}$. Noting that out of the nine elements in the system matrix, only two depend on the first dimension. They are the elements $S_{1,2}$ and $S_{3,1}$, and so $\mathfrak{H}_1^{1,2}$ and $\mathfrak{H}_1^{3,1}$ can be calculated according to (9.3) or (9.4). The rest is considered constant from the point of view of p_1. For the constant state space elements (i.e., $S_{1,1}$, $S_{1,3}$, $S_{2,1}$, $S_{2,2}$, $S_{3,3}$), their SVD-based column equivalent matrices $\mathfrak{H}_1^{o,i}$ are calculated as

$$\sqrt{\prod_{m=2\ldots3} M_m} \mathbf{1}_{M_1 \times 1} \left\| \mathcal{T}_{o,i} \right\|.$$

For the nondependent dimensions, $\mathfrak{H}_1^{2,3}$ and $\mathfrak{H}_1^{3,2}$ are calculated as

$$\mathfrak{H}_1^{2,3} = \mathbf{1}_{M_1 \times 1} \left\| \mathcal{T}_{2,3} \right\| \quad \text{and} \quad \mathfrak{H}_1^{3,2} = \sqrt{M_3} \mathbf{1}_{M_1 \times 1} \left\| \mathcal{T}_{3,2} \right\|,$$

respectively, according to Equation (9.5).

Calculating the HOSVD core tensor   The original approach to find the core tensor $\mathcal{D}$ is to use the already calculated $\mathbf{U}_n$ matrices and the discretized $\mathcal{S}^d$ tensor:

$$\mathcal{D} = \mathcal{S}^d \underset{n=1}{\overset{N}{\boxtimes}} \mathbf{U}_n^+. \tag{9.6}$$

In the modified method, we do not have the whole discretized tensor ($\mathcal{S}^d$), just the discretized subtensors ($\mathcal{T}_{o,i}$). This is not a problem since these tensors contain all the necessary information about the discretized system,

$$\mathcal{S}_{o,i}^d = \mathcal{T}_{o,i} \underset{n \notin J_{o,i}}{\boxtimes} \mathbf{1}_{M_n \times 1}. \tag{9.7}$$

The $\mathcal{D}$ core tensor can thus be generated element by element.

$$\mathcal{D}_{o,i} = \mathcal{S}_{o,i}^d \underset{n=1}{\overset{N}{\boxtimes}} \mathbf{U}_n^+ = \mathcal{T}_{o,i} \underset{n \in J_{o,i}}{\boxtimes} \mathbf{U}_n^+ \underset{n \notin J_{o,i}}{\boxtimes} \mathbf{U}_n^+ \mathbf{1}_{M_n \times 1} \tag{9.8}$$

Note that for the above only the $\mathcal{T}_{o,i}$ tensor and $\mathbf{U}_n$ matrices are needed, and so $\mathcal{S}_{o,i}^d$ is not computed explicitly. Note also that the algorithm may use the same $\mathbf{U}_n^+ \mathbf{1}_{M \times 1}$ vector for many (o, i) elements, so this quantity can be precalculated to reduce the complexity further.

Step 3: Identification of TP model components

After executing the first two steps, we get $\mathcal{D}$ and $\mathbf{U}_n$ matrices, which are equivalent to the results of the original method. The procedure for calculating the weight functions at any given point is thus the same in both the original and the modified TP model transformations.

9.3 Evaluation of computational reduction

This section describes the computational requirements of the original and the modified, computationally relaxed TP model transformations introduced above. Three classes of computational load will be evaluated and compared. The first concerns the discretization complexity of the algorithms, the second on the computational load caused by HOSVD, and the third evaluates the computational load of generating the tensor products. We here assume that the discretization grid is defined by M_n, $n = 1, \ldots, N$, and the size of the system matrix is $O \times I$.

9.3.1 Discretization complexity

The discretization complexity depends on the size of the processed data. In the original TP model transformation, all the elements of the system matrix $\mathbf{S}(\mathbf{p})$ are discretized for all dimensions of the parameter vector $\mathbf{p}$. The discretized tensor $\mathcal{S}^d$ thus has the size of $M_1 \times \cdots \times M_N \times O \times I$.

In the modified TP model transformation, the elements of the system matrix $\mathbf{S}(\mathbf{p})$ are discretized separately and only for the dimensions of $\mathbf{p}$ on which they depend. Thus, we do not have a huge tensor $\mathcal{S}^d$ that contains all discretization values but subtensors $\mathcal{T}_{o,i}$ that contain the most compact discretized subset. The total number of discretized elements is

$$\sum_{o=1}^{O} \sum_{i=1}^{I} \prod_{\forall j \in J_{o,i}} M_j.$$

It is obvious that if all the elements of $\mathbf{S}(\mathbf{p})$ depend on all dimensions of $\mathbf{p}$, no complexity reduction can be possible. In most cases, however, the modified TP model transformation does result in a great reduction on the number of discretized elements.

In a typical case, M should be about 100 in each dimension. With this setting, the previous example (9.2) requires a storage space of

$$9 \times M_1 \times M_2 \times M_3 = 9000000$$

numbers with the original method. By comparison, the same setting would require the storage of only

$$7 + M_2 + M_1 \times M_2 + M_2 \times M_3 + M_1 \times M_3 = 30107$$

numbers with the modified method.

9.3.2   HOSVD computation

Denote as $C_{\text{SVD}}(R)$ the computational load of conducting SVD on a matrix with R number of elements. As such, counting the number of elements in the matrix of concern and the number of SVD executions (N times), the original TP model transformation has a computation load of

$$C_{\text{original}} = N \cdot C_{\text{SVD}}(O \cdot I \cdot \prod_{n=1}^{N} M_n)$$

By comparison, while the modified TP model transformation executes the SVD many more times, the number of elements in the matrices involved are considerably fewer than the case of the original TP model transformation. This is an important fact given that the SVD operation needs about three to four times more memory space than that in storing the elements alone. In most cases, the bottleneck of the TP model transformation is actually in its memory consumption. The upper bound of the computational load of the modified TP transformation is given by

$$C_{\text{modified}} = \sum_{n=1}^{N} \left(\sum_{o=1}^{O} \sum_{i=1}^{I} C_{\text{SVD}} \left(\prod_{\forall j \in J_{o,i}} M_j \right) + C_{\text{SVD}} \left(M_n \cdot (O \cdot I \cdot M_n + 1) \right) \right)$$

In the extreme case, when all the elements of $S(\mathbf{p})$ depend on all dimensions of $\mathbf{p}$, the computational load of the modified TP model transformation is indeed higher than the original method because of the additional SVD operations. Even so, due to its much better memory consumption, the modified method can still be executed with the excess computation readily absorbed during the process. Of course, in the normal situations there will be many constant elements in the system matrix. Hence, the number of parameter-dependent elements is significantly fewer than $O \cdot I$ and also, typically, not all of them are dependent on all N dimensions of $\mathbf{p}$.

Applying the above to the example (9.2) again, the original method yields an HOSVD computational load of $C_{\text{original}} = 3C_{\text{SVD}}(9000000)$, while the modified method yields $C_{\text{modified}} = C_{\text{SVD}}(20100) + C_{\text{SVD}}(20200) + C_{\text{SVD}}(20100)$.

9.3.3 Tensor product computation

The original TP model transformation determines the core tensor via the product of the M^N-sized discretized tensor and N number of U_n^+ matrices having I_n rows and M columns. For simplicity, we assume that $K = I_1 = ... = I_N$. Therefore the computation of the tensor product requires the following multiplications:

$$C_{original} = O \cdot I \cdot \sum_{n=1}^{N} K^n \cdot M^{N+1-n}$$

The above takes into account that while executing a tensor product in each dimension, the size of that dimension is reduced from M to K. The modified TP model transformation uses Equation (9.8), and by a rough estimation yields:

$$C_{modified} \leq N \cdot K \cdot M + O \cdot I \cdot \left(\sum_{n=1}^{\gamma} K^n \cdot M^{\gamma+1-n} + \sum_{n=\gamma+1}^{N} K^n \right),$$

where γ is the maximum number of dependent parameters in the system matrix. The first term on the right hand side is the precalculation of the $U_n^+ 1_{M \times 1}$ vectors (which, technically speaking, does not require multiplications since it is just a summation). The two summations in the second term account for the tensor products in the dependent and nondependent dimensions, respectively, as according to (9.8). We can see again that a considerable reduction can be achieved if γ is less than the number of parameters.

Referring to the example (9.2) and assuming that $K = 5$ so the n-mode rank of the sampled tensor is 5 for each dimension n, the original tensor product yields

$$C_{original} = 9 \cdot \sum_{n=1}^{3} 5^n \cdot 100^{4-n} = 47362500,$$

which is about 47 million multiplications. By comparision, with $\gamma = 2$ in this case we obtain

$$C_{modified} \leq 3 \cdot 5 \cdots 100 + 3 \cdot 3 \cdot \left(\sum_{n=1}^{2} 5^n \cdot 100^{3-n} + \sum_{n=3}^{3} 5^n \right) = 479625,$$

a considerably smaller number. Actually, for this simple example we can calculate $C_{modified}$ exactly as

$$C_{modified} = 1500 + 5 \cdot 155 + 1 \cdot 650 + 3 \cdot 52625 = 160800$$

multiplications, which is even smaller. Here, the first term is the precalculation, and the remaining ones are the numbers of multiplications needed for the elements that are dependent on 0, 1, and 2 parameters in system (9.2).

9.4   Examples

9.4.1   A simple numerical example

To illustrate the advantages of the computationally reduced TP model transformation, let us consider a simple system first:

$$\mathbf{S}(p_1, p_2, p_3) = \begin{pmatrix} p_1 p_2^2 & p_2 p_3 \\ p_3 & 1 \end{pmatrix}$$

where the parameters p_1, p_2, and p_3 are all in the $[-1, 1]$ domain. To get reasonably smooth weighting functions and precise representation, let each parameter be sampled at $M = 50$ grid points.

Original TP model transformation: With the original method and sampling of the 2×2 system matrix over a $50 \times 50 \times 50$ grid set, roughly 4 *megabytes* of memory is needed to store the sampled data with double precision. To execute HOSVD, the discretized data should be laid out along each dimension. The original transformation is hence required to conduct SVD of a 50×100000 matrix for each parameter. The transformation took about 16 seconds on a regular PC.

Modified TP model transformation: The system matrix depends on three parameters, but none of its elements depend on all three. This means that the modified method will require fewer resources, as the samplings of $S_{1,1}(p_1, p_2), S_{1,2}(p_2, p_3), S_{2,1}(p_3), S_{2,2}$ elements are performed separately and only for their dependent parameters. The size of the sampled data is $50 \cdot 50 + 50 \cdot 50 + 50 + 1 = 5051$, so about 40 *kilobytes* of memory is needed in the modified case. As described in Section 9.2, HOSVD can be executed separately for each sampled element of the system matrix, which can then be readily reconstructed to yield the original HOSVD. The transformation takes less than a second on a regular PC.

9.4.2   The double inverted pendulum example

The goal of this example is to demonstrate that the modified, computationally relaxed TP model transformation can easily deal with complex and high-dimensional problems. We here select the model of the double inverted pendulum for illustration. We will show that execution of the original TP model transformation on this model requires tremendous computational power to carry out. By contrast, the modified TP model transformation as introduced can readily be executable on a standard PC to generate the desired HOSVD-based canonical form.

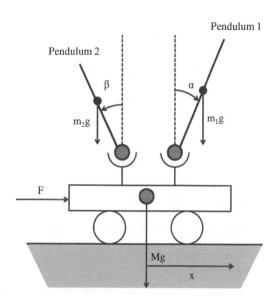

Figure 9.3: Parallel type double inverted pendulum system. (Modified from [NPBH09]–S. Nagy, Z. Petres, P. Baranyi, and H. Hashimoto, "Computational relaxed TP model transformation: restricting the computation to subspaces of the dynamic model," *Asian Journal of Control*, 11(5):461–475, 2009.)

Consider the parallel type double inverted pendulum system as shown in Figure 9.3 consisting of a straight line rail, a cart moving on the rail, a long Pendulum 1, a short Pendulum 2, and a driving unit. The parameters $M = 1.0kg$, $m_1 = 0.3kg$, and $m_2 = 0.1kg$ are the masses of the cart, Pendulum 1, and Pendulum 2, respectively. The parameter $g = 9.8m/s^2$ is the gravity acceleration. Suppose the mass of each pendulum is distributed uniformly. The center of mass of the long Pendulum 1 will be at the midpoint of its length located at $l_1 = 0.6m$, and that of Pendulum 2 at $l_2 = 0.2m$. The position of the cart from the rail origin is denoted as x, and is positive when the cart is on the right side of the rail origin. The angles of Pendulum 1 and Pendulum 2 from their upright positions are denoted as α and β, respectively, with clockwise direction being positive. The driving force applied horizontally to the cart is denoted as $F(N)$, with the right side as positive direction. We also assume that there is no friction in the pendulum system.

Equations of motion: The equation of motion in qLPV state space form is

$$\dot{\mathbf{x}}(t) = \mathbf{S}(\mathbf{p}(t)) \begin{pmatrix} \mathbf{x}(t) \\ u(t) \end{pmatrix} = \mathbf{A}(\mathbf{p}(t))\mathbf{x}(t) + \mathbf{B}(\mathbf{p}(t))u(t), \qquad (9.9)$$

$$A_1 \quad := \quad \left(1 - \frac{3}{4}\cos^2(\alpha)\right)m_1$$

$$A_2 \quad := \quad \left(1 - \frac{3}{4}\cos^2(\beta)\right)m_2 \tag{9.10}$$

$$A_3 \quad := \quad \frac{4}{3}(M + A_1 + A_2)$$

$$\mathbf{A}(\mathbf{p}(t)) = \begin{bmatrix} 0 & 0 & 0 & 1 & 0 & 0 \\ 0 & 0 & 0 & 0 & 1 & 0 \\ 0 & 0 & 0 & 0 & 0 & 1 \\ 0 & -\frac{m_1 g \sin(\alpha)\cos(\alpha)}{\alpha A_3} & -\frac{m_2 g \sin(\beta)\cos(\beta)}{\beta A_3} & 0 & \frac{4}{3}\frac{m_1 l_1 \dot{\alpha} \sin(\alpha)}{A_3} & \frac{4}{3}\frac{m_2 l_2 \dot{\beta} \sin(\beta)}{A_3} \\ 0 & \frac{(M+m_1+A_2)g\sin(\alpha)}{l_1 \alpha A_3} & \frac{3}{4}\frac{m_2 g \sin(\beta)\cos(\beta)\cos(\alpha)}{l_1 \beta A_3} & 0 & -\frac{m_1 \dot{\alpha}\sin(\alpha)\cos(\alpha)}{A_3} & -\frac{m_2 l_2 \dot{\beta}\cos(\alpha)\sin(\beta)}{l_1 A_3} \\ 0 & \frac{3}{4}\frac{m_1 g \sin(\alpha)\cos(\alpha)\cos(\beta)}{l_2 \alpha A_3} & \frac{(M+m_2+A_1)g\sin(\beta)}{l_2 \beta A_3} & 0 & -\frac{m_1 l_1 \dot{\alpha}\sin(\alpha)\cos(\beta)}{l_2 A_3} & -\frac{m_2 \dot{\beta}\sin(\beta)\cos(\beta)}{A_3} \end{bmatrix} \tag{9.11}$$

$$\mathbf{B}(\mathbf{p}(t)) = \begin{bmatrix} 0 & 0 & 0 & \frac{4}{3}\frac{1}{A_3} & -\frac{\cos(\alpha)}{l_1 A_3} & -\frac{\cos(\beta)}{l_2 A_3} \end{bmatrix}^T \tag{9.12}$$

where the dependent parameters, the state vector, and the control signal are

$$\mathbf{p}(t) = \begin{bmatrix} \alpha & \beta & \dot{\alpha} & \dot{\beta} \end{bmatrix}^T; \quad \mathbf{x}(t) = \begin{bmatrix} x & \alpha & \beta & \dot{x} & \dot{\alpha} & \dot{\beta} \end{bmatrix}^T; \quad u = F,$$

and the matrices $\mathbf{A}(\mathbf{p}(t))$ and $\mathbf{B}(\mathbf{p}(t))$ are defined in (9.11) and (9.12) with the simplifying notation given by (9.10). As such, the resulting qLPV model depends on *four* parameters. Let the transformation space be defined as

$$\Omega = \begin{bmatrix} -\frac{\pi}{6} & \frac{\pi}{6} \\ -\frac{\pi}{6} & \frac{\pi}{6} \\ -\pi & \pi \\ -\pi & \pi \end{bmatrix}.$$

So we are to perform the sampling in the ± 30 degree domain of the angles and the ± 180 degree/second domain of the angular velocities. The sampling grid is set to $M = 100$ for each parameter.

Original TP model transformation: Discretization of the original TP model transformation would result in a tensor of the size $100 \times 100 \times 100 \times 100 \times 6 \times 7$. This tensor thus contains $4.2 \cdot 10^9$ elements, which requires about 31.3 *gigabytes* of memory for double-precision computation. This means that the *discretization step* in the original TP model transformation alone already poses a strong memory requirement, not mentioning the execution of HOSVD, which would require considerably more space and computational capacity.

Modified TP model transformation: The 6×7-sized system matrix $\mathbf{A}(\mathbf{p})$ contains 15 parameter-dependent elements (see Equations 9.11 and 9.12). Of the 15, 9

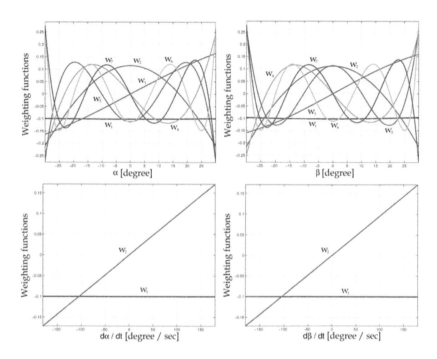

Figure 9.4: Weighting functions of the HOSVD-based canonical form.

depend on two parameters, 6 depend on three parameters, and none depends on all four parameters. So the size of the discretized data is $27 \cdot 1 + 9 \cdot 100^2 + 6 \cdot 100^3 \approx 6 \cdot 10^6$ elements, which needs about 46.5 *megabytes* of memory. Thus the modified TP model transformation can simply be executed on a standard PC of modest capacity. With the setup we have, the discretization step was performed in 4 minutes, and the HOSVD computation on the reduced tensors required only 8 seconds. The finite element TP model result is

$$\mathbf{S}(\mathbf{p}(t)) = \sum_{i=1}^{7}\sum_{j=1}^{7}\sum_{k=1}^{2}\sum_{l=1}^{2} w_{1,i}(\alpha)w_{2,j}(\beta)w_{3,k}(\dot{\alpha})w_{4,l}(\dot{\beta})\mathbf{D}_{i,j,k,l} = \mathcal{D} \overset{4}{\underset{n=1}{\boxtimes}}\ \mathbf{w}_n(p_n),$$

where the size of $\mathcal{D}$ is $7 \times 7 \times 2 \times 2 \times 6 \times 7$. This means that the numbers of weighting functions in the dimensions of $\mathbf{p}$ are $7, 7, 2, 2$, as shown in Figure 9.4. Note that in this example the weighting functions may not have closed analytical forms.

Part II

TP Model-Based Control System Design

Chapter 10

Overview of TP Model-Based Design Strategy

As mentioned, the tremendous advances in computer technology over the past two decades has supported a strong proliferation of computational-based control system design methodologies. The present book, to a large extent, is to tap into this new computation-based design paradigm of feedback, minimizing analytical derivation as much as possible. In Part I of this book, we introduced the algorithms to numerically generate a finite element tensor product (TP) type polytopic model, or TP model for short, for a given quasi-linear parameter-varying (qLPV) system. In Part II, we will describe the computation-based methodologies of control system design using the TP model representation. Then in Part III, a number of specific applications on the TP model control system design method will be given.

This chapter will briefly outline the three steps in TP model-based control system design strategy and present a simple example for the first step. Details of step two and step three will be given in the following two chapters. The strategy includes the following three steps.

First, we apply the TP model transformation to obtain the TP model from the given qLPV model:

$$\begin{pmatrix} \dot{\mathbf{x}} \\ \mathbf{y} \end{pmatrix} = \mathcal{S} \underset{n=1}{\overset{N}{\boxtimes}} \mathbf{w}_n(p_n) \begin{pmatrix} \mathbf{x} \\ \mathbf{u} \end{pmatrix}. \tag{10.1}$$

As mentioned, the TP model transformation is applicable regardless of whether the qLPV model is given in analytical form or other numeric and soft-computing algorithms. Note that we also have the freedom to shape the convexity of (10.1) using the various algorithms presented in Chapter 6.

The second step involves determining the feedback gains of the controller. In this regard, any suitable techniques can be applied to effectuate a design, if possible. We are aiming at, of course, computation-based methodologies in this book. We specifically adopt the use of linear matrix inequality (LMI) design under the parallel distributed compensation (PDC) framework for this purpose. The PDC framework was initially proposed by Tanaka and Wang in 1995 [WTG95] and later extended to handle multiobjective control design specifications in [TW01]. In particular, the structure of the PDC framework allows direct application of the TP model more readily than other types of LMI-based design formulations. It will be an efficient illustration of the TP model-based control design strategy. With experience and insight gained, readers are welcome to use other types of computation-based techniques for their design study.

The relevant LMI theorems under the PDC framework will be described in more detail in the next chapter.

The PDC framework calls for searching for a controller in the same polytopic form as the model:

$$\mathbf{u} = -\left(\mathcal{F} \overset{N}{\underset{n=1}{\boxtimes}} \mathbf{w}_n(p_n)\right)\mathbf{x}, \tag{10.2}$$

where the linear time-invariant (LTI) gains $\mathbf{F}_{i_1,i_2,\dots,i_N}$ stored in tensor $\mathcal{F}$ are called *vertex feedback gains*. Note that controller (10.2) shares the same structure as the model (10.1); each feedback gain corresponds to a LTI vertex system $\mathbf{S}_{i_1,i_2,\dots,i_N}$ of the tensor $\mathcal{S}$. The feedback gains are the outcomes of a multiobjective optimization process on a set of linear matrix inequalities (LMIs) formulated to reflect the desired control performance, as illustrated below.

$$\mathcal{S} \rightarrow \text{LMI+optimization parameters} \rightarrow \mathcal{F}. \tag{10.3}$$

The relevant LMI theorems under the PDC framework used to facilitate this design process will be described in detail in the next chapter.

Last but not least, focusing on formulating the set of LMIs in order to achieve a better controller design in the second step, which is primarily and almost exclusively the attention of LMI-based techniques in the literature today, may not be enough. The third step of the strategy calls for manipulating also the convex hull as defined by the convexity of the TP model. The algorithms introduced in Chapter 6 to incorporate the desired convexity into the ensuing TP model constitute the tools for manipulation. Such manipulation is necessary since the feasibility and performance of a LMI solution is very sensitive to the convex hull. LMI optimization actually gives the best solution for all the polytopic models contained in the convex hull, which may not be the best solution for the given qLPV problem. This is the reason why this technique is considered to be very conservative. Properly modifying the convex hull may result in better performance and decrease the conservativeness of the solution. This topic will be discussed in more detail in Chapter 12. Note that the second and third steps may need to be conducted iteratively in order to get the most satisfactory solution.

We will now give more details on the first step of the control design strategy and provide an example of a TP model ready for the start of the design process. In order to simplify the multiple notations and adopt the typical format utilized in the LMI theorems, let us use the linear equivalent indexing for the LTI components as

$$\mathbf{S}_r = \mathbf{S}_{i_1,i_2,\ldots,I_N},$$

where index $r = 1,\ldots,R$, $R = \prod_{n=1}^N I_n$, is the linear equivalent of the index $i_1, i_2, \ldots, i_N$ of the N-dimensional array of size $I_1 \times I_2 \times \ldots \times I_N$. The index r here is equivalent to the index h defined in (2.10).

The feedback gains $\mathbf{F}_r$ as derived from $\mathbf{S}_r$ based on LMI optimization can be restored to tensor $\mathcal{F}$ using the N-dimensional indexing as:

$$\mathbf{F}_{i_1,i_2,\ldots,I_N} = \mathbf{F}_r.$$

Define the multivariable weighting functions using the same linear indexing as above,

$$w_r(\mathbf{p}) = \prod_{n=1}^N w_{n,i_n}(p_n), \tag{10.4}$$

then we have the model:

$$\begin{pmatrix} \dot{\mathbf{x}} \\ \mathbf{y} \end{pmatrix} = \sum_{r=1}^R w_r(\mathbf{p})\mathbf{S}_r \begin{pmatrix} \mathbf{x} \\ \mathbf{u} \end{pmatrix} \tag{10.5}$$

and the controller:

$$\mathbf{u} = -\left(\sum_{r=1}^R w_r(\mathbf{p})\mathbf{F}_r\right)\mathbf{x}. \tag{10.6}$$

Note that only the indexing is changed here from that of (10.1) and (10.2). The LTI components, their weighting functions, and, hence, the convex hull stay the same.

As an example, let us use the translational oscillations with an eccentric rotational proof mass actuator (TORA) system derived in Sections 4.3 and 6.4 for illustration. In this case, $\mathbf{p} = [x_3\ x_4]^T$. We specifically consider TP model 2 in Section 6.4. We generate the two-variable weighting functions $w_r(x_3, x_4)$ by multiplying the corresponding one-variable weighting functions $w_{1,i_1}(x_3)$ and $w_{2,i_2}(x_4)$. The relations between r and i_1, i_2 are:

$$r = (i_1 - 1)I_1 + i_2, \quad i_1 = 1,\ldots,5, \quad i_2 = 1, 2. \tag{10.7}$$

TP model 2 of the TORA system then becomes:

$$\dot{\mathbf{x}}(t) = \sum_{r=1}^{10} w_r(x_3(t), x_4(t))\,(\mathbf{A}_r\mathbf{x}(t) + \mathbf{B}_r u(t)).$$

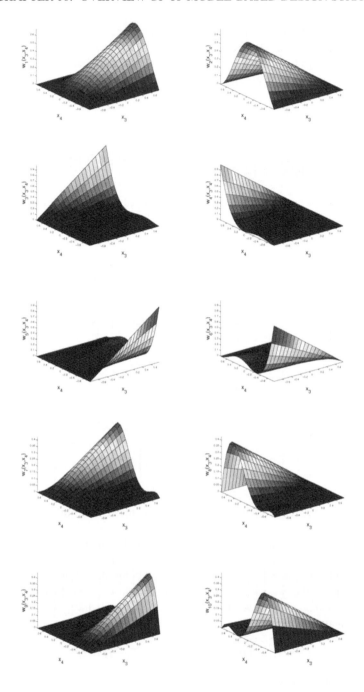

Figure 10.1: Individual two-variable weighting functions of TP model 2 over Ω.

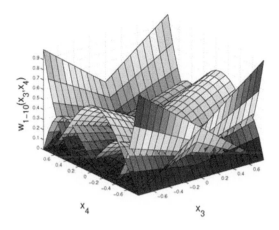

Figure 10.2: Plotting all two-variable weighting functions of TP model 2 together over Ω.

Figures 10.1 and 10.2 show the two-variable weighting functions. Figure 10.1 shows the individual $w_r(x_3(t), x_4(t))$ weighting functions and Figure 10.2 shows all of them together over the domain Ω in the $(x_3(t)\ x_4(t))$ plane. It can be observed that the LTI systems dominate over the border of Ω, and only one LTI system describes the dynamic model of the TORA at each corner of Ω. There is, however, no dominant LTI model in the equilibrium point. All of the LTI models have a significantly lower dominance compared to that at the border.

Chapter 11

LMI Theorems under the PDC Framework

In this chapter, we will first introduce the concept of linear matrix inequality (LMI) and then present some LMI theorems for control design under the parallel distributed compensation (PDC) framework.

11.1 LMIs for control system design

This section is based on the works of Scherer [SW00]. Linear matrix inequalities (LMIs) and LMI techniques have emerged as powerful design tools in areas ranging from control engineering to system identification and structural design. There are three factors that make LMI techniques appealing:

- A variety of design specifications and constraints can be expressed as LMIs.

- Once formulated in terms of LMIs, a problem can be exactly solved by efficient convex optimization algorithms (the "LMI solvers").

- While most problems with multiple constraints or objectives lack analytical solutions in terms of matrix equations, they often remain tractable in the LMI framework. This makes LMI-based design a valuable alternative to classical "analytical" methods.

The most significant advantage of LMIs is that it is easy to numerically specify and combine numerous design constraints, conditions, and goals in a tractable manner. Many control problems and design specifications have LMI formulations [BEGFB94]. This is especially true for Lyapunov-based analysis and design, and also for optimal LQG control, H_∞ control, minimal covariance control problems, and so forth. Further applications of LMIs arise in estimation, identification, optimal design, structural design, matrix scaling problems, and others. A non-exhaustive list of problems addressed by LMI techniques includes:

- Robust stability of systems with linear time-invariant (LTI) uncertainty (μ-analysis) [PD93, SD91, YND95]

- Quadratic stability of differential inclusions [BY89, HB76]

- Lyapunov stability of parameter-dependent systems [GAC94]

- Input/state/output properties of LTI systems (invariant ellipsoids, decay rate, etc.) [BEGFB94]

- Multimodel/multiobjective state feedback design [BSU93, Bar83, BEGFB94, CG96, KR91]

- Robust pole placement

- Optimal LQG control [BEGFB94]

- Robust H_∞ control [GA94, IS94]

- Multiobjective H_∞ synthesis [CG96, KR91, MOS98]

- Control of stochastic systems [BEGFB94]

- Weighted interpolation problems [BEGFB94]

11.1.1   Definition of LMIs

We now give the definition of an LMI-related concept often used in the literature, followed by the introduction of the LMIs.

Definition 11.1 (Affine function). A function $f : \mathcal{S} \mapsto \mathcal{T}$ is *affine* if $f(x) = f_0 + T(x)$ where $f_0 \in \mathcal{T}$ and $T : \mathcal{S} \mapsto \mathcal{T}$ is a linear map, that is,

$$T(\alpha_1 x_1 + \alpha_2 x_2) = \alpha_1 T(x_1) + \alpha_2 T(x_2)$$

for all $x_1, x_2 \in \mathcal{S}$ and $\alpha_1, \alpha_2 \in \mathbb{R}$.

Hence $f : \mathbb{R}^n \mapsto \mathbb{R}^m$ is affine only if there exists $x_0 \in \mathbb{R}^n$, such that the mapping $x \mapsto f(x) - f(x_0)$ is linear. This means that all affine functions $f : \mathbb{R}^n \mapsto \mathbb{R}^m$ can be represented as $f(x) = f(x_0) + T \cdot (x - x_0)$, where T is some matrix of dimension $m \times n$ and the dot $\cdot$ denotes multiplication. We will be interested in the case where $m = 1$. Furthermore, let us denote by $\langle \cdot, \cdot \rangle$ the standard inner product in $\mathbb{R}^n$, that is, for $x_1, x_2 \in \mathbb{R}^n$, $\langle x_1, x_2 \rangle = x_2^T x_1$.

A LMI is an expression of the form:

$$\mathbf{F}(\mathbf{x}) = \mathbf{F}_0 + \mathbf{x}_1 \mathbf{F}_1 + \cdots + \mathbf{x}_m \mathbf{F}_m > 0, \tag{11.1}$$

where

1. $\mathbf{x} = (\mathbf{x}_1, \ldots, \mathbf{x}_m)$ is a vector of m real numbers called the *decision variables*.

2. $\mathbf{F}_0, \ldots, \mathbf{F}_m$ are real symmetric matrices, that is, $\mathbf{F}_i = \mathbf{F}_i^T \in \mathbb{R}^{n \times n}, i = 0, \ldots, m$ for some $n \in \mathbb{Z}_+$.

3. The inequality > 0 in (11.1) means "positive definite," that is, $\mathbf{u}^T \mathbf{F}(\mathbf{x}) \mathbf{u} > 0$ for all $\mathbf{u} \in \mathbb{R}^n, \mathbf{u} \neq 0$. Equivalently, the smallest eigenvalue of $\mathbf{F}(\mathbf{x})$ is positive.

In more general terms, we have

Definition 11.2 (Linear Matrix Inequality). A *linear matrix inequality* (LMI) is an inequality

$$\mathbf{F}(\mathbf{x}) > 0, \tag{11.2}$$

where $\mathbf{F}$ is an *affine function* mapping a finite dimensional vector space $\mathbb{V}$ to the set $\mathbb{S} := \{M \mid \exists n > 0 \text{ such that } \mathbf{M} = \mathbf{M}^T \in \mathbb{R}^{n \times n}\}$ of real symmetric matrices.

Remark 11.1. An affine mapping $\mathbf{F} : \mathbb{V} \mapsto \mathbb{S}$ necessarily takes the form $\mathbf{F}(\mathbf{x}) = \mathbf{F}_0 + \mathbf{T}(\mathbf{x})$ where $\mathbf{F}_0 \in \mathbb{S}$ and $\mathbf{T} : \mathbb{V} \mapsto \mathbb{S}$ is a linear transformation. Thus if $\mathbb{V}$ is finite dimensional, say of dimension m, and $\{\mathbf{e}_1, \ldots, \mathbf{e}_m\}$ constitutes a basis for $\mathbb{V}$, then we can write

$$\mathbf{T}(\mathbf{x}) = \sum_{j=1}^{m} \mathbf{x}_j \mathbf{F}_j,$$

where the elements $\{\mathbf{x}_1, \ldots, \mathbf{x}_m\}$ are such that $\mathbf{x} = \sum_{j=1}^{m} \mathbf{x}_j \mathbf{e}_j$ and $\mathbf{F}_j = \mathbf{T}(\mathbf{e}_j)$ for $j = 1, \ldots, m$. Hence we obtain (11.1) as a special case.

Remark 11.2. The same remark applies to affine mappings $\mathbf{F} : \mathbb{R}^{n \times n} \mapsto \mathbb{S}$. A simple example is the Lyapunov inequality $\mathbf{F}(\mathbf{X}) = \mathbf{A}^T \mathbf{X} + \mathbf{X} \mathbf{A} + \mathbf{Q} > 0$. Here, $\mathbf{A}, \mathbf{Q} \in \mathbb{R}^{n \times n}$ are assumed to be given and $\mathbf{X} \in \mathbb{R}^{n \times n}$ is unknown. The unknown variable is therefore a *matrix*. Note that this defines an LMI only if $\mathbf{Q}$ is symmetric. We can view this LMI as a special case of (11.1) by defining a basis $\mathbf{E}_1, \ldots, \mathbf{E}_m$ of $\mathbb{R}^{n \times n}$ and writing $\mathbf{X} = \sum_{j=1}^{m} \mathbf{x}_j \mathbf{E}_j$. Indeed,

$$\mathbf{F}(\mathbf{X}) = \mathbf{F}\left(\sum_{j=1}^{m} \mathbf{x}_j \mathbf{E}_j\right) = \mathbf{F}_0 + \sum_{j=1}^{m} \mathbf{x}_j \mathbf{F}(\mathbf{E}_j) = \mathbf{F}_0 + \sum_{j=1}^{m} \mathbf{x}_j \mathbf{F}_j,$$

which is of the form (11.1).

Remark 11.3. A *nonstrict* LMI is a LMI where $>$ in (11.1) and (11.2) is replaced by $\geq$.

Remark 11.4. The matrix inequalities $\mathbf{F}(\mathbf{x}) < 0$ and $\mathbf{F}(\mathbf{x}) > \mathbf{G}(\mathbf{x})$ with $\mathbf{F}$ and $\mathbf{G}$ affine functions are obtained as special cases of Definition 11.2 as they can be rewritten as the linear matrix inequality $-\mathbf{F}(\mathbf{x}) > 0$ and $\mathbf{F}(\mathbf{x}) - \mathbf{G}(\mathbf{x}) > 0$.

11.1.2   Constraints expressed via LMIs

The LMI (11.2) defines a *convex constraint* on $\mathbf{x}$. That is, the set $\mathcal{S} := \{\mathbf{x} \mid \mathbf{F}(\mathbf{x}) > 0\}$ is convex. Indeed, if $\mathbf{x}_1, \mathbf{x}_2 \in \mathcal{S}$ and $\alpha \in (0, 1)$ then

$$\mathbf{F}(\alpha \mathbf{x}_1 + (1 - \alpha)\mathbf{x}_2) = \alpha \mathbf{F}(\mathbf{x}_1) + (1 - \alpha)\mathbf{F}(\mathbf{x}_2) > 0,$$

where in the equality $\mathbf{F}$ is affine, and the inequality follows from the fact that $\alpha \geq 0$ and $(1 - \alpha) \geq 0$.

Although the convex constraint $\mathbf{F}(\mathbf{x}) > 0$ on $\mathbf{x}$ may seem rather special, it transpires that many convex sets can be represented in this way. In the following we discuss some seemingly trivial properties of LMIs that are of eminent help in the reduction of multiple constraints on an unknown variable to an equivalent constraint involving a single LMI.

Definition 11.3 (System of Linear Matrix Inequalities). A *system of Linear Matrix Inequalities* is a finite set of LMIs

$$\mathbf{F}_1(\mathbf{x}) > 0, \mathbf{F}_2(\mathbf{x}) > 0, \ldots, \mathbf{F}_k(\mathbf{x}) > 0. \tag{11.3}$$

It is a simple but essential property that every system of LMIs can be rewritten as one single LMI. Specifically, $\mathbf{F}_1(\mathbf{x}) > 0, \mathbf{F}_2(\mathbf{x}) > 0, \ldots, \mathbf{F}_k(\mathbf{x}) > 0$ if and only if

$$\mathbf{F}(\mathbf{x}) = \begin{pmatrix} \mathbf{F}_1(\mathbf{x}) & 0 & , \ldots & 0 \\ 0 & \mathbf{F}_2(\mathbf{x}) & \ldots & 0 \\ \vdots & & \ddots & \vdots \\ 0 & 0 & \ldots & \mathbf{F}_k(\mathbf{x}) \end{pmatrix} > 0.$$

The last inequality indeed makes sense as $\mathbf{F}(\mathbf{x})$ is symmetric for any $\mathbf{x}$. Furthermore, since the set of eigenvalues of $\mathbf{F}(\mathbf{x})$ is simply the union of the eigenvalues of $\mathbf{F}_1(\mathbf{x}), \ldots, \mathbf{F}_k(\mathbf{x})$, any $\mathbf{x}$ that satisfies $\mathbf{F}(\mathbf{x}) > 0$ also satisfies the system of LMIs (11.3) and vice versa.

A second important property amounts to incorporating *affine constraints* in LMIs. By this, we mean that *combined constraints* (in the unknown $\mathbf{x}$) of the form

$$\begin{cases} \mathbf{F}(\mathbf{x}) > 0 \\ \mathbf{C}\mathbf{x} = \mathbf{d} \end{cases}$$

or

$$\begin{cases} \mathbf{F(x)} > 0 \\ \mathbf{x} = \mathbf{Ay} + \mathbf{b} \text{ for some } \mathbf{y} \in \mathbb{R}^n. \end{cases}$$

where the affine function $\mathbf{F} : \mathbb{R}^m \mapsto \mathbb{S}$ and matrices $\mathbf{A} \in \mathbb{R}^{m \times n}$ and $\mathbf{b} \in \mathbb{R}^m$ are given, can be *lumped* in one single LMI. More generally, the combined equations

$$\begin{cases} \mathbf{F(x)} > 0 \\ \mathbf{x} \in \mathcal{M}, \end{cases} \tag{11.4}$$

where $\mathcal{M}$ is an *affine subset* of $\mathbb{R}^m$, that is,

$$\mathcal{M} = \mathbf{x}_0 + \mathcal{M}_0 = \{\mathbf{x}_0 + \mathbf{m}_0 \mid \mathbf{m}_0 \in \mathcal{M}_0\},$$

with $\mathbf{x}_0 \in \mathbb{R}^m$ and $\mathcal{M}_0$ a linear subspace of $\mathbb{R}^m$, can be rewritten in the form of a single LMI. To actually see this, let $\mathbf{e}_1, \ldots, \mathbf{e}_{m_0} \in \mathbb{R}^m$ be a basis of $\mathcal{M}_0$ and let $\mathbf{F(x)} = \mathbf{F}_0 + \mathbf{T(x)}$ be decomposed as in Remark 11.1. Then (11.4) can be rewritten as

$$\begin{aligned} 0 < \mathbf{F(x)} &= \mathbf{F}_0 + \mathbf{T}\left(\mathbf{x}_0 + \sum_{j=1}^{m_0} x_j \mathbf{e}_j\right) = \underbrace{\mathbf{F}_0 + \mathbf{T(x_0)}}_{\text{constant part}} + \underbrace{\sum_{j=1}^{m_0} x_j \mathbf{T(e}_j)}_{\text{linear part}} \\ &= \overline{\mathbf{F}}_0 + x_1 \overline{\mathbf{F}}_1 + \cdots + x_{m_0} \overline{\mathbf{F}}_{m_0} \\ &= \overline{\mathbf{F}}(\overline{\mathbf{x}}), \end{aligned}$$

where $\overline{\mathbf{F}}_0 = \mathbf{F}_0 + \mathbf{T(x_0)}, \overline{\mathbf{F}}_j = \mathbf{T(e}_j)$, and $\overline{\mathbf{x}} = (x_1, \ldots, x_{m_0})$ are the coefficients of $\mathbf{x} - \mathbf{x}_0$ in the basis of $\mathcal{M}_0$. This implies that $\mathbf{x} \in \mathbb{R}^m$ satisfies (11.4) if and only if $\overline{\mathbf{F}}(\overline{\mathbf{x}}) > 0$. Note that the dimension m_0 of $\overline{\mathbf{x}}$ is smaller than the dimension m of $\mathbf{x}$.

A third property of LMIs is obtained from a simple exercise in algebra. It is possible to convert some *nonlinear* inequalities to *linear inequalities*. Suppose that we partition a matrix $\mathbf{M} \in \mathbb{R}^{n \times n}$ as

$$\mathbf{M} = \begin{pmatrix} \mathbf{M}_{11} & \mathbf{M}_{12} \\ \mathbf{M}_{21} & \mathbf{M}_{22} \end{pmatrix},$$

where $\mathbf{M}_{11}$ has dimension $r \times r$. Assume that $\mathbf{M}_{11}$ is nonsingular. Then the matrix $\mathbf{S} = \mathbf{M}_{22} - \mathbf{M}_{21} \mathbf{M}_{11}^{-1} \mathbf{M}_{12}$ is called the *Schur complement* of $\mathbf{M}_{11}$ in $\mathbf{M}$. If $\mathbf{M}$ is symmetric then we have

$$\begin{aligned} \mathbf{M} > 0 &\iff \begin{pmatrix} \mathbf{M}_{11} & 0 \\ 0 & \mathbf{S} \end{pmatrix} > 0 \\ &\iff \begin{cases} \mathbf{M}_{11} > 0 \\ \mathbf{S} > 0 \end{cases}. \end{aligned}$$

In conclusion, we can state the following proposition:

Proposition 11.1 (Schur complement). *Let* $\mathbf{F} : \mathbb{V} \mapsto \mathbb{S}$ *be an affine function which is partitioned according to*

$$\mathbf{F}(\mathbf{x}) = \begin{pmatrix} \mathbf{F}_{11}(\mathbf{x}) & \mathbf{F}_{12}(\mathbf{x}) \\ \mathbf{F}_{21}(\mathbf{x}) & \mathbf{F}_{22}(\mathbf{x}) \end{pmatrix},$$

where $\mathbf{F}_{11}(\mathbf{x})$ *is square. Then* $\mathbf{F}(\mathbf{x}) > 0$ *if and only if*

$$\begin{cases} \mathbf{F}_{11}(\mathbf{x}) > 0 \\ \mathbf{F}_{22}(\mathbf{x}) - \mathbf{F}_{12}(\mathbf{x}) \, [\mathbf{F}_{11}(\mathbf{x})]^{-1} \, \mathbf{F}_{21}(\mathbf{x}) > 0. \end{cases} \tag{11.5}$$

Note that the second inequality in (11.4) is a *nonlinear* matrix inequality in $\mathbf{x}$. Using this result, it follows that nonlinear matrix inequalities of the form (11.5) can be converted to LMIs. In particular, it follows that nonlinear inequalities of the form (11.5) define a convex constraint on the variable $\mathbf{x}$ in the sense that all $\mathbf{x}$ values satisfying (11.5) define a convex set.

11.1.3   Generic problems for LMIs

The publications listed in Section 11.1 show that many optimization problems in control, identification, and signal processing can be formulated (or reformulated) using LMIs. Clearly, it only makes sense to cast these problems in an LMI setting if these inequalities can be solved in an efficient and reliable way. Since the LMI $\mathbf{F}(\mathbf{x}) > 0$ defines a *convex constraint* on the variable $\mathbf{x}$, optimization problems involving the minimization (or maximization) of a performance function $f : \mathcal{S} \mapsto \mathbb{R}$ with $\mathcal{S} := \{\mathbf{x} \mid \mathbf{F}(\mathbf{x}) > 0\}$ belong to the class of *convex optimization problems*. From the perspective of the previous section, it is apparent that the full power of convex optimization theory can be employed if the performance function f is known to be convex.

Suppose that $\mathbf{F}, \mathbf{G}, \mathbf{H} : \mathbb{V} \mapsto \mathbb{S}$ are affine functions. There are three generic problems related to LMIs:

1. **Feasibility:** The test whether or not there exist solutions $\mathbf{x} \in \mathbb{V}$ of $\mathbf{F}(\mathbf{x}) > 0$ is called a *feasibility problem*. The LMI is called *feasible* if such an $\mathbf{x}$ exists; otherwise the LMI $\mathbf{F}(\mathbf{x}) > 0$ is said to be *infeasible*.

2. **Optimization:** Let $f : \mathcal{S} \mapsto \mathbb{R}$ and suppose that $\mathcal{S} = \{\mathbf{x} \mid \mathbf{F}(\mathbf{x}) > 0\}$. The problem to determine

$$V_{\text{opt}} = \inf_{\mathbf{x} \in \mathcal{S}} f(\mathbf{x})$$

is called an *optimization problem with an LMI constraint*. This problem involves the determination of V_{opt} and for arbitrary $\varepsilon > 0$ the calculation of an *almost optimal solution* $\mathbf{x}$ which satisfies $\mathbf{x} \in \mathcal{S}$ and $V_{\text{opt}} \leq f(\mathbf{x}) \leq V_{\text{opt}} + \varepsilon$.

3. **Generalized eigenvalue problem:** The generalized eigenvalue problem amounts to minimizing a scalar $\lambda \in \mathbb{R}$ subject to

$$\begin{cases} \lambda \mathbf{F}(\mathbf{x}) - \mathbf{G}(\mathbf{x}) > 0 \\ \mathbf{F}(\mathbf{x}) > 0 \\ \mathbf{H}(\mathbf{x}) > 0 \end{cases} .$$

A major breakthrough in convex optimization lies in the introduction of interior point methods. These methods were developed in a series of papers [Kar84a], [Kar84b] and applied to the context of LMI problems by Yurii Nesterov and Arkadii Nemirovskii [NN94]. The interior point method is the most popular optimization technique thanks to its efficiency. It is also widely used for solving LMIs in commercial scientific applications such as MATLAB Optimization Toolbox and Robust Control Design Toolbox [GNLC95]. Interested readers are referred to the above cited works for detailed descriptions of the mathematical approach involved.

11.2 LMI optimization under the PDC framework

The topic of LMI-based control design is a very active research area. There are numerous formulations in the literature, where one can readily find very relevant and powerful LMIs for one's desired control specifications. As mentioned earlier, in this book we will focus on the PDC framework for LMI-based control design. The intent of this section is to introduce a few basic LMI theorems under the PDC context which will be used later in the control examples. It will be shown in the next chapter that these simple basic LMI techniques can even be more powerful if we properly manipulate the convex hull defined by the given tensor product (TP) model.

We would like to point out that though we will be using the continuous case in the illustration examples of our design approach, the PDC design framework can actually be applied to discrete polytopic models with the following form as well:

Definition 11.4 (Discrete convex polytopic model).

$$\begin{pmatrix} \mathbf{x}(t+1) \\ \mathbf{y}(t) \end{pmatrix} = \sum_{r=1}^{R} w_r(\mathbf{p}(t)) \mathbf{S}_r \begin{pmatrix} \mathbf{x}(t) \\ \mathbf{u}(t) \end{pmatrix}, \tag{11.6}$$

where the weighting functions and vector $\mathbf{p}(t) \in \Omega$ are as defined in the continuous case.

11.2.1   Lyapunov stability criteria

We can apply Lyapunov stability theory to derive the following stability criteria for open-loop systems. The proof can be found in [TS90], [TS92].

Assume that we have a quasi-linear parameter-varying (qLPV) model

$$\begin{pmatrix} \dot{\mathbf{x}} \\ \mathbf{y} \end{pmatrix} = \sum_{r=1}^{R} w_r(\mathbf{p}) \mathbf{S}_r \begin{pmatrix} \mathbf{x} \\ \mathbf{u} \end{pmatrix}, \tag{11.7}$$

and we search the controller in the form of:

$$\mathbf{u} = -\left(\sum_{r=1}^{R} w_r(\mathbf{p}) \mathbf{F}_r \right) \mathbf{x}. \tag{11.8}$$

Theorem 11.1 (Stability of continuous convex polytopic model). *The equilibrium of the continuous polytopic model (11.7) with* $\mathbf{u}(t) = 0$ *is globally asymptotically stable if there exists a common positive definite matrix* $\mathbf{P}$ *such that*

$$\mathbf{A}_r^T \mathbf{P} + \mathbf{P} \mathbf{A}_r < 0, \quad r = 1, \dots, R, \tag{11.9}$$

that is, a common $\mathbf{P}$ *has to exist for all* $\mathbf{A}_r$.

Theorem 11.2 (Stability of discrete convex polytopic model). *The equilibrium of the discrete polytopic model (11.6) with* $\mathbf{u}(t) = 0$ *is globally asymptotically stable if there exists a common positive definite matrix* $\mathbf{P}$ *such that*

$$\mathbf{A}_r^T \mathbf{P} \mathbf{A}_r - \mathbf{P} < 0, \quad r = 1, \dots, R, \tag{11.10}$$

that is, a common $\mathbf{P}$ *has to exist for all* $\mathbf{A}_r$.

Let us derive the stability criteria for closed-loop systems. When we substitute the controller for the model we obtain the following:

$$\dot{\mathbf{x}}(t) = \sum_{r=1}^{R} \sum_{s=1}^{R} w_r(\mathbf{p}(t)) w_s(\mathbf{p}(t)) \left(\mathbf{A}_r - \mathbf{B}_r \mathbf{F}_s \right) \mathbf{x}(t)$$

and

$$\mathbf{x}(t+1) = \sum_{r=1}^{R} \sum_{s=1}^{R} w_r(\mathbf{p}(t)) w_s(\mathbf{p}(t)) \left(\mathbf{A}_r - \mathbf{B}_r \mathbf{F}_s \right) \mathbf{x}(t),$$

for the continuous system and the discrete system, respectively. Denote $\mathbf{G}_{r,s} = \mathbf{A}_r - \mathbf{B}_r \mathbf{F}_s$. Then one can write, respectively,

$$\dot{\mathbf{x}}(t) = \sum_{r=1}^{R} w_r(\mathbf{p}(t)) w_r(\mathbf{p}(t)) \mathbf{G}_{r,r} \mathbf{x}(t) + 2 \sum_{r=1}^{R} \sum_{s>r}^{R} w_r(\mathbf{p}(t)) w_s(\mathbf{p}(t)) \left(\frac{\mathbf{G}_{r,s} + \mathbf{G}_{s,r}}{2} \right) \mathbf{x}(t)$$

and

$$\mathbf{x}(t+1) = \sum_{r=1}^{R} w_r(\mathbf{p}(t))w_r(\mathbf{p}(t))\mathbf{G}_{r,r}\mathbf{x}(t) + 2\sum_{r=1}^{R}\sum_{s>r}^{R} w_r(\mathbf{p}(t))w_s(\mathbf{p}(t))\left(\frac{\mathbf{G}_{r,s}+\mathbf{G}_{s,r}}{2}\right)\mathbf{x}(t).$$

Applying Theorems 11.1 and 11.2, we have the following.

Theorem 11.3 (Stability for continuous convex polytopic model). *The equilibrium of the continuous system (11.7) with control value (11.8) is globally asymptotically stable if there exists a common positive definite matrix* $\mathbf{P}$ *such that for all* $r, s = 1, ..., R$:

$$\mathbf{G}_{r,r}^T\mathbf{P} + \mathbf{P}\mathbf{G}_{r,r} < 0,$$

and $\forall r < s$ *except* $\forall \mathbf{p}(t) : w_r(\mathbf{p}(t))w_s(\mathbf{p}(t)) = 0$:

$$\left(\frac{\mathbf{G}_{r,s}+\mathbf{G}_{s,r}}{2}\right)^T\mathbf{P} + \mathbf{P}\left(\frac{\mathbf{G}_{r,s}+\mathbf{G}_{s,r}}{2}\right) \leq 0.$$

Theorem 11.4 (Stability for discrete convex polytopic model). *The equilibrium of the discrete system (11.6) with control value (11.8) is globally asymptotically stable if there exists a common positive definite matrix* $\mathbf{P}$ *such that* $r, s = 1, ..., R$:

$$\mathbf{G}_{r,r}^T\mathbf{P}\mathbf{G}_{r,r} - \mathbf{P} < 0,$$

and $\forall r < s$ *except* $\forall \mathbf{p}(t) : w_r(\mathbf{p}(t))w_s(\mathbf{p}(t)) = 0$:

$$\left(\frac{\mathbf{G}_{r,s}+\mathbf{G}_{s,r}}{2}\right)^T\mathbf{P}\left(\frac{\mathbf{G}_{r,s}+\mathbf{G}_{s,r}}{2}\right) - \mathbf{P} \leq 0.$$

11.2.2 Control design for stability

Finding the feedback gains which satisfy the stability criteria given in Theorems 11.3 and 11.4 may not be simple. However, if we can reformulate these theorems into LMIs then we can use numerical methods to solve them. Since the examples given later in this book all have continuous qLPV models, we focus on LMI-based stability design theorems developed for continuous models from now on. Note that with a simple modification, and based on the theorems given for discrete systems in the previous section, we can readily derive LMIs for the discrete systems as well. These derivations and other LMIs developed for multiobjective control design of discrete systems are detailed in [TW01].

The conditions in stability Theorems 11.3 and 11.4 are not jointly convex in $\mathbf{F}_r$ and $\mathbf{P}$. To convert them, let us multiply the inequalities on the left and right by $\mathbf{P}^{-1}$ and define new variable $\mathbf{X} = \mathbf{P}^{-1}$. Then, define $\mathbf{M}_r = \mathbf{F}_r\mathbf{X}$ so that for $\mathbf{X} > 0$ we have $\mathbf{F}_r = \mathbf{M}_r\mathbf{X}^{-1}$. We thus arrive at the following theorems:

Theorem 11.5 (Asymptotic stability design for continuous convex polytopic models). *The polytopic model (11.7) with control value (11.8) is asymptotically stable if there exist* $\mathbf{X} > 0$ *and* $\mathbf{M}_r$ *satisfying equations*

$$-\mathbf{X}\mathbf{A}_r^T - \mathbf{A}_r\mathbf{X} + \mathbf{M}_r^T\mathbf{B}_r^T + \mathbf{B}_r\mathbf{M}_r > 0 \qquad (11.11)$$

for all r and

$$-\mathbf{X}\mathbf{A}_r^T - \mathbf{A}_r\mathbf{X} - \mathbf{X}\mathbf{A}_s^T - \mathbf{A}_s\mathbf{X} + \mathbf{M}_s^T\mathbf{B}_r^T + \mathbf{B}_r\mathbf{M}_s + \mathbf{M}_r^T\mathbf{B}_s^T + \mathbf{B}_s\mathbf{M}_r \geq 0 \qquad (11.12)$$

for $r < s \leq R$, *except the pairs* (r, s) *such that* $\forall \mathbf{p}(t) : w_r(\mathbf{p}(t))w_s(\mathbf{p}(t)) = 0$, *where the feedback gains are determined from the solutions* $\mathbf{X}$ *and* $\mathbf{M}_r$ *as*

$$\mathbf{F}_r = \mathbf{M}_r\mathbf{X}^{-1}. \qquad (11.13)$$

The LMIs of this theorem can be readily solved by mathematical programs, for example, see MATLAB LMI Control Toolbox [GNLC95]. It is worth emphasizing again that the design theorem introduced above is one of the first published LMI theorems under the PDC framework. Deriving more relaxed LMIs is an ongoing research area. In the related literature we can find a number of alternative and more powerful LMIs. One may search for more recently published design theorems if those offered here are not feasible for the given control design problem.

11.2.3   Multiobjective control optimization

We will now present several theorems on applying the LMIs developed under the PDC framework to multiobjective control design. Detailed derivation of these theorems is given in [TW01]. They serve to prepare for the introduction of convex hull manipulation in the next chapter. It will be shown that the LMIs in these theorems are appropriate for a large class of control problems (see later examples), and their performance is considerably enhanced via convex hull manipulation. Readers may also wish to search recent literature for possibly more powerful LMIs to apply.

One major concern in control system is the speed of the response, for example, the decay rate as given by the largest Lyapunov exponent. We have the following theorem:

Theorem 11.6 (Decay rate control). *Assume we have the polytopic model (11.7) with the controller (11.8). The largest lower bound on the decay rate by the quadratic Lyapunov function is guaranteed by the solution of the following generalized eigenvalue minimization problem (GEVP):*

$$\underset{\mathbf{X},\mathbf{M}_1,...,\mathbf{M}_R}{\text{maximize}} \quad \alpha \quad \text{subject to}$$

$$\mathbf{X} > \mathbf{0},$$
$$-\mathbf{X}\mathbf{A}_r^T - \mathbf{A}_r\mathbf{X} + \mathbf{M}_r^T\mathbf{B}_r^T + \mathbf{B}_r\mathbf{M}_r - 2\alpha\mathbf{X} > \mathbf{0},$$
$$-\mathbf{X}\mathbf{A}_r^T - \mathbf{A}_r\mathbf{X} - \mathbf{X}\mathbf{A}_s^T - \mathbf{A}_s\mathbf{X} + \mathbf{M}_s^T\mathbf{B}_r^T + \mathbf{B}_r\mathbf{M}_s$$
$$+\mathbf{M}_r^T\mathbf{B}_s^T + \mathbf{B}_s\mathbf{M}_r - 4\alpha\mathbf{X} \geq \mathbf{0},$$

for $r < s \leq R$, except the pairs (r, s) such that $\forall \mathbf{p}(t) : w_r(\mathbf{p}(t))w_s(\mathbf{p}(t)) = 0$. In this case, the feedback gains are determined from the GEVP solutions using (11.13).

In practical control designs we also have to deal with the physical constraints of the system. The theorem below ensures that such constraint is satisfied via the embedded LMIs.

Theorem 11.7 (Constraint on the control value). *Assume that $\|\mathbf{x}(0)\| \leq \phi$, where $\mathbf{x}(0)$ is unknown, but the upper bound ϕ is known. The constraint $\|\mathbf{u}(t)\|_2 \leq \mu$ is enforced at all times $t \geq 0$ if the LMIs*

$$\phi^2\mathbf{I} \leq \mathbf{X}$$
$$\begin{pmatrix} \mathbf{X} & \mathbf{M}_r^T \\ \mathbf{M}_r & \mu^2\mathbf{I} \end{pmatrix} \geq \mathbf{0}$$

hold.

Theorem 11.8 (Constraint on the output). *Assume that $\|\mathbf{x}(0)\| \leq \phi$, where $\mathbf{x}(0)$ is unknown, but the upper bound ϕ is known. The constraint $\|\mathbf{y}(t)\|_2 \leq \lambda$ is enforced at all times $t \geq 0$ if the LMIs*

$$\phi^2\mathbf{I} \leq \mathbf{X}$$
$$\begin{pmatrix} \mathbf{X} & \mathbf{X}\mathbf{C}_r^T \\ \mathbf{C}_r\mathbf{X} & \lambda^2\mathbf{I} \end{pmatrix} \geq \mathbf{0}$$

hold.

In order to ensure the above condition for a large set of initial states, we can set ϕ to be a large quantity even if $x(0)$ is unknown. However, one should note that a large ϕ could lead to conservative designs.

Note that the LMIs of Theorems 11.7 and 11.8 must be simultaneously solved with the LMIs of the selected stability theorem. LMIs for additional constraints, such as disturbance rejection, for both continuous and discrete systems can be found in [TW01].

11.2.4 Simultaneous observer/controller design

Previous sections are dedicated to the design of state feedback controllers. In practical applications, however, the state of the system is often not readily available. In this case we may try to determine the state from the system response to some input over some period of time. Therefore, we need to utilize output feedback design that is based on the simultaneous designs of a state feedback controller, and an observer capable of estimating the unavailable state values from the output of the system.

Noting that there are various alternative methods of output feedback and observer design available (e.g., see [SW00], [TW01]), we first of all give the polytopic observer structure we are going to adopt in this book. The observers are required to satisfy

$$\mathbf{x}(t) - \hat{\mathbf{x}}(t) \to 0 \quad \text{as} \quad t \to \infty,$$

where $\hat{\mathbf{x}}(t)$ denotes the state vector estimated by the observer. This guarantees that the steady-state error between $\mathbf{x}(t)$ and $\hat{\mathbf{x}}(t)$ converges to zero. In order to achieve this goal, we introduce the following observer structure:

$$
\begin{aligned}
\dot{\hat{\mathbf{x}}} &= \mathbf{A}(\mathbf{p})\hat{\mathbf{x}} + \mathbf{B}(\mathbf{p})\mathbf{u} + \mathbf{K}(\mathbf{p})(\mathbf{y} - \hat{\mathbf{y}}) \\
\hat{\mathbf{y}} &= \mathbf{C}(\mathbf{p})\hat{\mathbf{x}}.
\end{aligned}
$$

The polytopic model form of this structure is

$$\dot{\hat{\mathbf{x}}} = \sum_{r=1}^{R} \mathbf{w}_r(\mathbf{p})\mathbf{A}_r\hat{\mathbf{x}} + \sum_{r=1}^{R} \mathbf{w}_r(\mathbf{p})\mathbf{B}_r\mathbf{u} + \sum_{r=1}^{R} \mathbf{w}_r(\mathbf{p})\mathbf{K}_r(\mathbf{y} - \hat{\mathbf{y}}) \qquad (11.14)$$

$$\hat{\mathbf{y}} = \sum_{r=1}^{R} \mathbf{w}_r(\mathbf{p})\mathbf{C}_r\hat{\mathbf{x}}.$$

The goal of the observer design is to determine gains $\mathbf{K}_r, r = 1, \ldots, R$, in such a way that the stability of the observer and the controller is simultaneously guaranteed, namely, the stability of the output feedback control structure is guaranteed.

For simplicity, we assume that $\mathbf{p}$ does not contain values from the estimated state vector $\hat{\mathbf{x}}$. For those cases when the vector $\mathbf{p}$ contains values from $\hat{\mathbf{x}}$ we refer to [TW01]. The vertex gains $\mathbf{K}_r$ of the observer and the feedback gains $\mathbf{F}_r$ of the controller can be derived, for instance, by the feasibility test of the following LMIs:

Theorem 11.9 (Globally and asymptotically stable observer and controller). *Assume we have the polytopic model (11.7) with controller (11.8) and observer structure (11.14). This output feedback control structure is globally and asymptotically stable if there exists such $\mathbf{P}_1 > 0, \mathbf{P}_2 > 0$ and $\mathbf{M}_{1,r}, \mathbf{N}_{2,r}$ ($r = 1, \ldots, R$) satisfying equations*

$$
\begin{aligned}
\mathbf{P}_1\mathbf{A}_r^T - \mathbf{M}_{1,r}^T\mathbf{B}_r^T + \mathbf{A}_r\mathbf{P}_1 - \mathbf{B}_r\mathbf{M}_{1,r} &< \mathbf{0}, \\
\mathbf{A}_r^T\mathbf{P}_2 - \mathbf{C}_r^T\mathbf{N}_{2,r}^T + \mathbf{P}_2\mathbf{A}_r - \mathbf{N}_{2,r}\mathbf{C}_r &< \mathbf{0}, \\
\mathbf{P}_1\mathbf{A}_r^T - \mathbf{M}_{1,s}^T\mathbf{B}_r^T + \mathbf{A}_s\mathbf{P}_1 - \mathbf{B}_r\mathbf{M}_{1,s} + \mathbf{P}_1\mathbf{A}_s^T - \mathbf{M}_{1,r}^T\mathbf{B}_s^T + \mathbf{A}_s\mathbf{P}_1 - \mathbf{B}_s\mathbf{M}_{1,r} &< \mathbf{0}, \\
\mathbf{A}_r^T\mathbf{P}_2 - \mathbf{C}_s^T\mathbf{N}_{2,r}^T + \mathbf{P}_2\mathbf{A}_r - \mathbf{N}_{2,r}\mathbf{C}_s + \mathbf{A}_s^T\mathbf{P}_2 - \mathbf{C}_r^T\mathbf{N}_{2,s}^T + \mathbf{P}_2\mathbf{A}_s - \mathbf{N}_{2,s}\mathbf{C}_r &< \mathbf{0},
\end{aligned}
$$

for $r < s \leq R$, except the pairs (r, s) such that $\forall \mathbf{p}(t) : w_r(\mathbf{p}(t))w_s(\mathbf{p}(t)) = 0$, and where $\mathbf{M}_{1,r} = \mathbf{F}_r\mathbf{P}_1$ and $\mathbf{N}_{2,r} = \mathbf{P}_2\mathbf{K}_r$. The feedback gains and the observer gains can then be obtained from the solution of the above LMIs as $\mathbf{F}_r = \mathbf{M}_{1,r}\mathbf{P}_1^{-1}$ and $\mathbf{K}_r = \mathbf{P}_2^{-1}\mathbf{N}_{2,r}$.

We can add additional LMIs to the above theorem to guarantee the various constraints as in the case of the controller design detailed in the previous section. Solving all the LMIs simultaneously thus leads to the desired controller.

Additional LMI theorems related to the design of controllers and observers and various other less restrictive LMI theorems can be found in [TW01].

11.3 TP model-based control design procedures

This section summarizes the previous discussion and gives the main steps of the TP model transformation-based design methodology:

Assume that the model is given in qLPV state space form as:

$$\begin{align}
\dot{\mathbf{x}}(t) &= \mathbf{A}(\mathbf{p}(t))\mathbf{x}(t) + \mathbf{B}(\mathbf{p}(t))\mathbf{u}(t), \tag{11.15}\\
\mathbf{y}(t) &= \mathbf{C}(\mathbf{p}(t))\mathbf{x}(t) + \mathbf{D}(\mathbf{p}(t))\mathbf{u}(t),
\end{align}$$

with input $\mathbf{u}(t)$, output $\mathbf{y}(t)$, and state vector $\mathbf{x}(t)$. The system matrix

$$\mathbf{S}(\mathbf{p}(t)) = \begin{pmatrix} \mathbf{A}(\mathbf{p}(t)) & \mathbf{B}(\mathbf{p}(t)) \\ \mathbf{C}(\mathbf{p}(t)) & \mathbf{D}(\mathbf{p}(t)) \end{pmatrix} \in \mathbb{R}^{O \times I} \tag{11.16}$$

Again, it is emphasized that the matrix function $\mathbf{S}(\mathbf{p}(t))$ in the present TP model-based design methodology can be given by analytical equations, by soft computing-based identification techniques, for example, by linguistic fuzzy rules, by a neural network defined by a huge array of connection weights and a graph specifying the connections, by genetic algorithms, or even by a black-box simulation program, for example, the MATLAB-Simulink Toolbox, or a combination of all of the above. The only requirement is that the given $\mathbf{S}(\mathbf{p}(t))$ must be discretizable, namely, we can compute the elements of $\mathbf{S}(\mathbf{p}(t))$ over certain vectors $\mathbf{p}(t)$ defined by the discretization grid.

The main steps of the TP model transformation-based control design methodology are provided below.

Step 1

Define the space of Ω. This is the space of the parameter vector $\mathbf{p}(t)$ usually defined by the physical constraints of the system to be controlled. It is worth pointing out that Ω can be conservatively defined first and fine-tuned later when we have a better idea of the specific values of $\mathbf{p}(t)$.

Step 2

Execute the TP model transformation on the qLPV model to yield the corresponding finite element high-order singular value decomposition (HOSVD)-based canonical form, and hence the finite element convex TP model:

$$\begin{pmatrix} \dot{\mathbf{x}}(t) \\ \mathbf{y}(t) \end{pmatrix} = \mathcal{S} \underset{n=1}{\overset{N}{\boxtimes}} \mathbf{w}_n(p_n(t)) \begin{pmatrix} \mathbf{x}(t) \\ \mathbf{u}(t) \end{pmatrix}. \tag{11.17}$$

Without better knowledge, we suggest adopting close-to-normality (CNO) weighting functions to yield a tight convex hull for (11.17) as a first approach. If the finite element TP form does not exist, then a trade-off between the number of LTI components and the approximation error will need to be conducted. We then need to check the numerical error between the resulting TP model and the original linear parameter-varying (LPV) model.

Step 3

Select LMI theorems according to the required multiobjective control performances. Then substitute the LTI components $\mathcal{S}$ obtained from Step 2 in the LMIs. By solving the LMIs we obtain the feedback gains $\mathcal{F}$ (and also the gains for the observer if that is the case). This immediately leads to the controller as

$$\mathbf{u}(t) = \left(-\mathcal{F} \underset{n=1}{\overset{N}{\boxtimes}} \mathbf{w}_n(p_n(t)) \right) \mathbf{x}(t), \tag{11.18}$$

or expressed in terms of multivariable weighting functions:

$$\mathbf{u}(t) = \left(-\sum_{r=1}^{R} w_r(\mathbf{p}(t)) \mathbf{F}_r \right) \mathbf{x}(t). \tag{11.19}$$

If the selected LMIs are infeasible then we can define other types of convex hull via the weighting functions in Step 2, or select other less conservative LMI-based theorems.

11.4   LMI-based control design for the TORA example

Let us continue the example of the translational oscillations with an eccentric rotational proof mass actuator (TORA) system studied in Sections 4.3 and 6.4. A detailed study is given in [PBKH07]. For the finite element convex TP model representation needed in Step 2 of the control design methodology above, we utilize the TP model 2 of the TORA system derived in Section 6.4:

$$\dot{\mathbf{x}}(t) = \mathbf{S}(\mathbf{p}(t))\begin{pmatrix}\mathbf{x}(t)\\u(t)\end{pmatrix} = \sum_{i=1}^{5}\sum_{j=1}^{2} w_{1,i}(x_3(t))w_{2,j}(x_4(t))\Big(\mathbf{A}_{i,j}\mathbf{x}(t) + \mathbf{B}_{i,j}u(t)\Big),$$

where CNO-type weighting functions have been adopted to yield a tight convex hull for the TORA model. The LTI systems are given by:

$$\mathbf{A}_{1,1} = \begin{pmatrix} 0 & 1.0000 & 0 & 0 \\ -1.0493 & 0 & 0 & 0.0094 \\ 0 & 0 & 0 & 1.0000 \\ 0.2288 & 0 & 0 & -0.0010 \end{pmatrix} \quad \mathbf{B}_{1,1} = \begin{pmatrix} 0 \\ -0.2288 \\ 0 \\ 1.0493 \end{pmatrix}$$

$$\mathbf{A}_{1,2} = \begin{pmatrix} 0 & 1.0000 & 0 & 0 \\ -1.0493 & 0 & 0 & -0.0094 \\ 0 & 0 & 0 & 1.0000 \\ 0.2288 & 0 & 0 & 0.0010 \end{pmatrix} \quad \mathbf{B}_{1,2} = \begin{pmatrix} 0 \\ -0.2288 \\ 0 \\ 1.0493 \end{pmatrix}$$

$$\mathbf{A}_{2,1} = \begin{pmatrix} 0 & 1.0000 & 0 & 0 \\ -1.0204 & 0 & 0 & 0.1133 \\ 0 & 0 & 0 & 1.0000 \\ 0.1443 & 0 & 0 & -0.0160 \end{pmatrix} \quad \mathbf{B}_{2,1} = \begin{pmatrix} 0 \\ -0.1443 \\ 0 \\ 1.0204 \end{pmatrix}$$

$$\mathbf{A}_{2,2} = \begin{pmatrix} 0 & 1.0000 & 0 & 0 \\ -1.0204 & 0 & 0 & -0.1133 \\ 0 & 0 & 0 & 1.0000 \\ 0.1443 & 0 & 0 & 0.0160 \end{pmatrix} \quad \mathbf{B}_{2,2} = \begin{pmatrix} 0 \\ -0.1443 \\ 0 \\ 1.0204 \end{pmatrix}$$

$$\mathbf{A}_{3,1} = \begin{pmatrix} 0 & 1.0000 & 0 & 0 \\ -1.0204 & 0 & 0 & -0.1133 \\ 0 & 0 & 0 & 1.0000 \\ 0.1443 & 0 & 0 & 0.0160 \end{pmatrix} \quad \mathbf{B}_{3,1} = \begin{pmatrix} 0 \\ -0.1443 \\ 0 \\ 1.0204 \end{pmatrix}$$

$$\mathbf{A}_{3,2} = \begin{pmatrix} 0 & 1.0000 & 0 & 0 \\ -1.0204 & 0 & 0 & 0.1133 \\ 0 & 0 & 0 & 1.0000 \\ 0.1443 & 0 & 0 & -0.0160 \end{pmatrix} \quad \mathbf{B}_{3,2} = \begin{pmatrix} 0 \\ -0.1443 \\ 0 \\ 1.0204 \end{pmatrix}$$

$$\mathbf{A}_{4,1} = \begin{pmatrix} 0 & 1.0000 & 0 & 0 \\ -1.0328 & 0 & 0 & 0.1439 \\ 0 & 0 & 0 & 1.0000 \\ 0.1884 & 0 & 0 & -0.0272 \end{pmatrix} \quad \mathbf{B}_{4,1} = \begin{pmatrix} 0 \\ -0.1884 \\ 0 \\ 1.0328 \end{pmatrix}$$

$$\mathbf{A}_{4,2} = \begin{pmatrix} 0 & 1.0000 & 0 & 0 \\ -1.0328 & 0 & 0 & -0.1439 \\ 0 & 0 & 0 & 1.0000 \\ 0.1884 & 0 & 0 & 0.0272 \end{pmatrix} \quad \mathbf{B}_{4,2} = \begin{pmatrix} 0 \\ -0.1884 \\ 0 \\ 1.0328 \end{pmatrix}$$

$$\mathbf{A}_{5,1} = \begin{pmatrix} 0 & 1.0000 & 0 & 0 \\ -1.0222 & 0 & 0 & -0.1578 \\ 0 & 0 & 0 & 1.0000 \\ 0.1572 & 0 & 0 & 0.0282 \end{pmatrix} \quad \mathbf{B}_{5,1} = \begin{pmatrix} 0 \\ -0.1572 \\ 0 \\ 1.0222 \end{pmatrix}$$

$$\mathbf{A}_{5,2} = \begin{pmatrix} 0 & 1.0000 & 0 & 0 \\ -1.0222 & 0 & 0 & 0.1578 \\ 0 & 0 & 0 & 1.0000 \\ 0.1572 & 0 & 0 & -0.0282 \end{pmatrix} \quad \mathbf{B}_{5,2} = \begin{pmatrix} 0 \\ -0.1572 \\ 0 \\ 1.0222 \end{pmatrix}$$

11.4.1   Control specifications

We summarize the typical design specifications of the TORA benchmark problem outlined in the special issue of [Ber98, page 309] and papers by [Tad01], [TTW98], [EOSR99]:

- The closed-loop system is stable.

- The closed-loop system exhibits good settling behavior for a class of initial conditions.

- The physical configuration of the system necessitates the constraint $|x_1| \leq 0.025\ m$, where x_1 is the translational position.

- The control torque value is limited by $N \leq 0.100\ Nm$, although somewhat higher torques can be tolerated for short periods.

The quantities x_1 and N here are as defined in Section 4.3.

11.4.2   State feedback control design

This section derives controllers for different specifications by applying the TP model 2 and the LMI stability theorems developed under the PDC framework (see Section 11.2). The control value is computed by (11.18) as:

$$u(t) = -\left(\sum_{i=1}^{5} \sum_{j=1}^{2} w_{1,i}(x_3(t)) w_{2,j}(x_4(t)) \mathbf{F}_{i,j} \right) \mathbf{x}(t).$$

Compared to the analytical solutions in the TORA benchmark papers mentioned above, an important difference is that here the solution is automatically derived via numerical methods in a few minutes on a regular computer without analytical interaction. If the dynamic model is modified then the design process can be repeated in a few minutes. This is unlike the analytical solutions, which may be a tedious process, even in the case of a small modification.

In order to satisfy all the constraints defined in Section 11.4.1, the LMIs of Theorems 11.7 and 11.8 are added to the LMIs of Theorem 11.5. We find that these LMIs are feasible. Figure 11.1 depicts the closed-loop performance of the translational position, rotational position, and control input for the initial condition of $\mathbf{x}(0) = \begin{pmatrix} 0.023\ 0\ 0\ 0 \end{pmatrix}$. Note that the solution achieved asymptotic stability, not merely the stability requirement as in the design specifications.

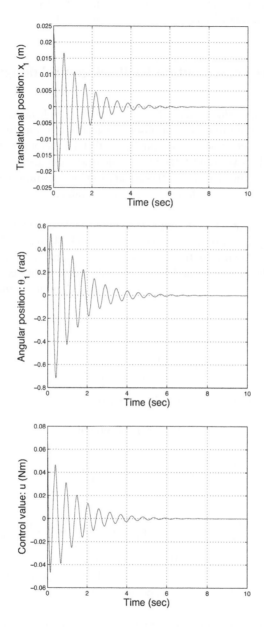

Figure 11.1: Results of the asymptotically stable controller with constraints $\mathbf{x}(t = 0) = (0.023\ 0\ 0\ 0)$.

In the above design we applied very simple LMIs. Additional design specifications such as parameter uncertainty, and so forth, can be readily considered by incorporating the relevant LMIs selected from recent literature.

11.4.3   Observer-based output feedback control design

Let us assume that state values $x_1(t)$ and $x_2(t)$ are not observable in the TORA system, and our goal is therefore to derive an observer capable of estimating their values. Before starting the observer design, we emphasize that the vector $\mathbf{p}(t)$ in this example does not contain values $x_1(t)$ and $x_2(t)$, in agreement with our assumption previously set in Section 11.2.4. The element $p_1(t)$ is equal to angular position $x_3(t)$ and $p_2(t)$ is equal to angular speed $x_4(t)$, both of which are measurable.

The controller and observer design is carried out as described in Section 11.2.4. We derive the LTI gains $\mathbf{K}$ of the observer structure (11.14) and the LTI gains $\mathbf{F}$ of the controller via Theorem 11.9. In order to satisfy the control specifications, we also add the LMIs of Theorems 11.7 and 11.8 to include the necessary constraints in the design. For the observer structure we define the matrix $\mathbf{C}$ for all r values from

$$\mathbf{y}(t) = \mathbf{Cx}(t),$$

as:

$$\mathbf{C}_r = \begin{pmatrix} 0 & 0 & 1 & 0 \\ 0 & 0 & 0 & 1 \end{pmatrix}.$$

The state values x_1 and x_2 are estimated by (11.14) as:

$$\hat{\mathbf{x}}(t) = \mathbf{A}(\mathbf{p}(t))\hat{\mathbf{x}}(t) + \mathbf{B}(\mathbf{p}(t))u(t)+$$

$$\left(\sum_{i=1}^{5} \sum_{j=1}^{2} w_{1,i}(x_3(t))w_{2,j}(x_4(t))\mathbf{K}_{i,j} \right) (\mathbf{y}(t) - \hat{\mathbf{y}}(t)),$$

where

$$\mathbf{y}(t) = \begin{pmatrix} x_3(t) \\ x_4(t) \end{pmatrix} \quad \text{and} \quad \hat{\mathbf{y}}(t) = \begin{pmatrix} \hat{x}_3(t) \\ \hat{x}_4(t) \end{pmatrix} \quad \text{and} \quad \mathbf{p}(t) = \begin{pmatrix} x_3(t) \\ x_4(t) \end{pmatrix}.$$

It can be confirmed that all the LMIs of the above theorems developed for the observer and the controller guaranteeing the constraints are feasible and an observer-based output feedback control design is possible in this case.

Figure 11.2 shows the simulated closed-loop system performance for the initial condition $\mathbf{x}(t = 0) = \begin{pmatrix} 0.023 & 0 & 0 & 0 \end{pmatrix}$ and initial observer state $\hat{\mathbf{x}}(t = 0) = \begin{pmatrix} 0 & 0 & 0 & 0 \end{pmatrix}$. The system is asymptotically stabilized. Compared with Figure 11.1, we can see that the stabilization time is slightly shorter without the observer than with the observer. On the other hand, the control torque magnitude is significantly reduced from a value of

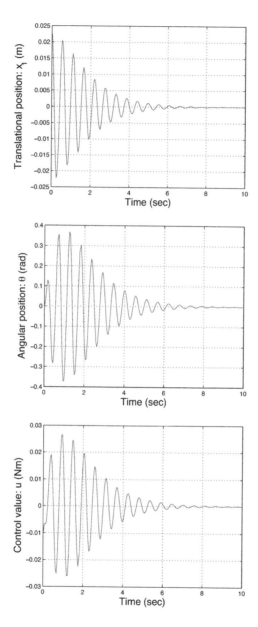

Figure 11.2: Results of the observer-based controller with constraints $\mathbf{x}(t = 0) = (0.023\ 0\ 0\ 0)$ and $\hat{\mathbf{x}}(t = 0) = (0\ 0\ 0\ 0)$.

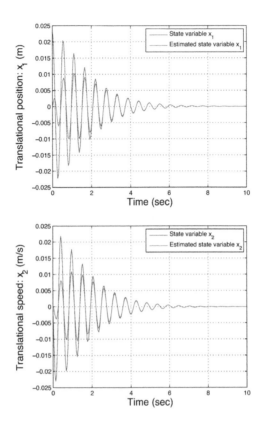

Figure 11.3: Convergence of the observer state variables to the actual system state variables.

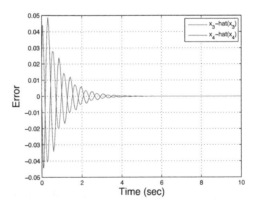

Figure 11.4: Convergence of the corresponding output error $(\mathbf{y} - \hat{\mathbf{y}})$ to zero.

roughly 0.06 *Nm* without to roughly 0.03 *Nm* with the observer. Furthermore, Figure 11.3 shows the convergence of the estimated state variables $\hat{x}_1(t)$ and $\hat{x}_2(t)$ to the state variables $x_1(t)$ and $x_2(t)$, while Figure 11.4 shows the error $\mathbf{y}(t) - \hat{\mathbf{y}}(t)$ converging to zero.

Chapter 12

Convex Hull Manipulation

In the prevailing literature, one can find a considerable number of papers on tractable and systematic manipulation of linear matrix inequalities (LMIs) aimed at achieving the best control performance. Most of these works are devoted to the derivation of new and more powerful LMIs to facilitate enhanced control system design. In contrast, relatively little attention has been given to the transformation of given models into proper polytopic, affine model representations in order to influence the feasibility and effectiveness of the applied LMIs, and hence the resulting control performance. In this regard, we note that recent papers have actually shown that different constructions of the system matrix in the quasi-linear parameter-varying (qLPV) model will lead to considerably different LMI solutions [HL96]. The purpose of this section is thus to show that the optimization of the control performance *must include* the manipulation of the convex hull in addition to the construction of LMIs, and that the tensor product (TP) model transformation as described here in this book offers a systematic solution.

This chapter focuses on three facts:

1. The LMIs optimize the desired performance within the convex hull defined by the linear time-invariant (LTI) vertex systems of the TP model representation.

2. The TP model of a given qLPV model is not unique. Different TP models can represent the same qLPV model while defining different convex hulls through their LTI vertex systems.

3. LMI optimization is a nonlinear transformation process highly sensitive on the convex hull upon which it is carried out. In case of only a slight modification of the convex hull, the LMI design may yield a very different solution or become infeasible.

We can conclude from the above that the LMI-based design yields an optimized solution for the given convex hull, rather than for the given qLPV problem. For the same qLPV model, if a different TP model defining a different convex hull is selected, the control result may be very different. This is true even if we utilize the same LMIs optimizing the same control performances. Therefore, when optimizing the desired control performance, we have to check which convex hull leads to the best solution of the applied LMIs. From another perspective, it would be meaningless to compare the effectiveness of different LMIs in different examples if the convex hull or set of vertex systems upon which they are based is not the same. Both the LMI and the convex hull manipulation must be investigated simultaneously for control system design. This is a new paradigm introduced in this book.

12.1   Nonlinear sensitivity of control solutions

Chapter 11 described the methodology to derive control systems for polytopic models via LMIs. As long as a given polytopic model is convex, which is the only requirement, exact details in the convex combination of its LTI vertex systems played no role in the controller derivation. The LMIs actually optimize the desired control performance for all models spanned by the different convex combinations of the LTI vertex systems of the given TP model. Based on this understanding, we present the following discussions.

Assume that we have a qLPV model. We generate a convex type TP model representation of the system matrix as:

$$\mathbf{S}(\mathbf{p}) = \mathcal{S} \underset{n=1}{\overset{N}{\boxtimes}} \mathbf{w}_n(p_n). \tag{12.1}$$

Then we properly select certain LMIs for our control objectives and determine the controller. The LMIs optimize our objectives and get as close to the desired control performance as possible, and the control is expressed as:

$$u = -\mathcal{F} \underset{n=1}{\overset{N}{\boxtimes}} \mathbf{w}_n(p_n)\mathbf{x}. \tag{12.2}$$

However, the TP model representation of a qLPV model is not invariant. Actually, we can generate an infinite number of very different convex type TP model representations for it. For instance, let us say we generate $z = 1, \ldots, Z$ different representations for the same qLPV model as:

$$\mathbf{S}(\mathbf{p}) = \mathcal{S}_z \underset{n=1}{\overset{N}{\boxtimes}} \mathbf{w}_{n,z}(p_n) \tag{12.3}$$

and apply the same LMIs to these Z TP models,

$$\mathcal{S}_z \underset{LMI}{\Rightarrow} \mathcal{F}_z. \tag{12.4}$$

We recognize that the LMI is a strongly nonlinear mapping from $\mathcal{S}_z$ to $\mathcal{F}_z$, which means that the solutions $\mathcal{F}_z$ are very different. We then obtain Z different controllers $\mathcal{F}_z$ using the same TP structure as $\mathcal{S}_z$:

$$u_z = -\mathcal{F}_z \overset{N}{\underset{n=1}{\boxtimes}} \mathbf{w}_{n,z}(p_n)\mathbf{x}. \tag{12.5}$$

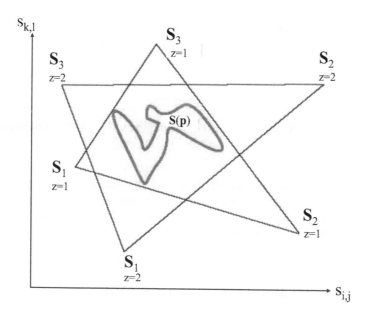

Figure 12.1: Two different TP model representations of the same qLPV model.

Figures 12.1 and 12.2 illustrate such a situation for the case of $z = 2$ and $N = 3$. Figure 12.1 shows the convex hulls generated by the N vertex systems (namely, $\mathbf{S}_1$, $\mathbf{S}_2$, and $\mathbf{S}_3$) of $\mathcal{S}_1$ and $\mathcal{S}_2$. They both embed the same qLPV model $\mathbf{S}(\mathbf{p}(t))$ depicted as the curved line in the middle (refer to Figure 6.1). Figure 12.2, on the other hand, shows the two convex hulls in the controller space generated by the N vertex feedback gains (namely, $\mathbf{F}_1$, $\mathbf{F}_2$, and $\mathbf{F}_3$) of $\mathcal{F}_1$ and $\mathcal{F}_2$ while sharing the same TP model structure as $\mathcal{S}_1$ and $\mathcal{S}_2$, respectively. It can be observed that the resulting controllers for the two cases, as depicted by the curves inside the convex hulls, are completely different. Hence, even the same LMIs applied to the same qLPV models can lead to completely different controllers and different performance. This means that choosing the most appropriate TP model representation is extremely important besides the LMI optimization in the controller design process.

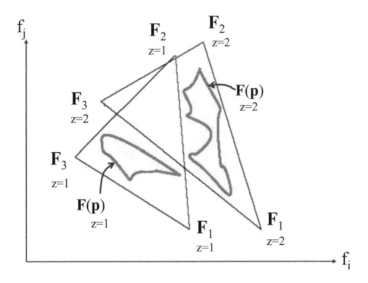

Figure 12.2: Controllers derived for the two different TP model representations of the same qLPV model (of Figure 12.1) by applying the same LMIs.

In conclusion, even when the optimality of all solutions is guaranteed by the LMIs, one has to investigate which one of the solutions is optimal overall. Such investigation should focus on how to generate different S_z for efficient evaluation of the best possible optimal performance.

12.2   Conservativeness of control solutions

The above can also be viewed from a different perspective. Assume again that we have a qLPV model with TP model representation:

$$\mathbf{S}(\mathbf{p}) = S \overset{N}{\underset{n=1}{\boxtimes}} \mathbf{w}_n(p_n), \tag{12.6}$$

and a controller from a LMI-based design:

$$u = -\mathcal{F} \overset{N}{\underset{n=1}{\boxtimes}} \mathbf{w}_n(p_n)\mathbf{x}. \tag{12.7}$$

As the LMI design process

$$S \underset{LMI}{\Rightarrow} \mathcal{F} \tag{12.8}$$

is independent of the weightings of the convex combination to form $\mathbf{S}(\mathbf{p})$, we are actually solving the same performance optimization problem for all qLPV models lying within the convex hull defined by the LTI vertex systems in S of our qLPV model. Specifically, we may define Q different additional sets of weighting functions $\mathbf{w}_{n,q}(p_n)$, $q = 1, \ldots, Q$, each guaranteeing the convexity, and use them to yield Q different models over the same LTI vertices S:

$$S_q(\mathbf{p}) = S \underset{n=1}{\overset{N}{\boxtimes}} \mathbf{w}_{n,q}(p_n). \tag{12.9}$$

Then the design process (12.8) will yield the same controller gain $\mathcal{F}$ for these models $S_q(\mathbf{p})$, $q = 1, \ldots, Q$. Thus we obtain Q number of different controllers:

$$u_q = -\mathcal{F} \underset{n=1}{\overset{N}{\boxtimes}} \mathbf{w}_{n,q}(p_n)\mathbf{x}. \tag{12.10}$$

The feasibility of the LMI-based design hence depends not only on our model $S(\mathbf{p})$ in (12.6) but also on the additional models $S_q(\mathbf{p})$ defined by the weighting functions $\mathbf{w}_{n,q}(p_n)$ within the same convex hull. In summary, the LMI optimizes the control solution not only for our model but for all models within the convex hull spanned by all possible sets of weightings functions.

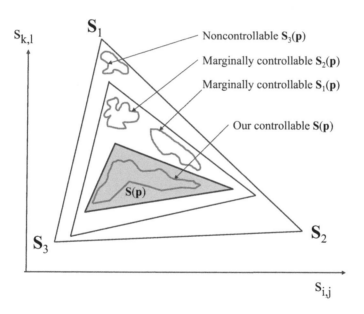

Figure 12.3: Different convex hulls for the same qLPV model yield different LMI-based designs.

The situation depicted in Figure 12.3 is for the case $Q = 3$. Even if the original weighting functions of (12.6) correspond to a readily controllable $S(\mathbf{p})$, should the convex hull be so large as to include a noncontrollable $S_3(\mathbf{p})$, then the LMI-based design will produce an infeasible outcome. Should the convex hull be smaller to include marginally controllable $S_1(\mathbf{p})$ and $S_2(\mathbf{p})$, then the LMI-based process will produce a marginal design. Only when the convex hull is tight enough in bounding our model $S(p)$ while excluding the noncontrollable ones will the LMI-based process produce a feasible design.

In general, we can conclude that if one of the models $S_q(p)$ in Figure 12.3 is not solvable or is unstable, then the LMI design for that convex hull is not feasible irrespective of whether our original model $S_q(p)$ is solvable or not. Furthermore, if our model is readily controllable, but one of the models $S_q(p)$ within the same convex hull is nearly uncontrollable, then, according to the given control performances involved in the LMIs, the solution of the LMIs generates a control value that involves unnecessarily high strength, in order to force all of the possible models—including the near uncontrollable one—to achieve the desired control requirements and constraints. This is the reason why the parallel distributed compensation (PDC) and the other similar LMI-based polytopic design strategies are considered to be conservative. It is therefore important to define the convex hull in such a way that it includes our model, but excludes the problematic models that we do not intend to control. Such manipulation can effectively reduce the conservativeness of the polytopic LMI-based design in our proposed strategy.

Part III

Control Design Examples

Chapter 13

Control Design with *TPtool* Toolbox

Part III of the book contains several practical examples and case studies to illustrate the applications of the tensor product (TP) model design strategy. These applications include a two-dimensional prototypical aeroelastic wing section in Chapter 14, and a three-dimensional version in Chapter 15 to demonstrate the readily extension of the TP model formulation to include higher complexity in the system model. A three degrees of freedom (3-DoF) helicopter with four propellers and a heavy vehicle rollover prevention problem are also treated in Chapter 16 and Chapter 17. If readers are interested, other TP model transformation-based control examples can be found in the following papers: [BY06], [SGNB08], [SHQW12], [RW11], [GWGL11], [QZL$^+$11], [CW09], [Chu10],[LPK12], [LKPV11], [AHJM12], [RSV11], [SGB10], [PPU$^+$08], [PDP$^+$12], [PH11], [IMK11], [PDP$^+$10], [IKM11], [KP06], [KPP06], [KSSK07], [TOP11].

Before we embark on the actual examples, we present a brief introduction of the control design commands in *TPtool* here. This complements the content of Chapter 7, which presents the algorithms for obtaining the TP model from a quasi-linear parameter-varying (qLPV) system.

The toolbox assumes linear matrix inequality (LMI)-based controller design using the parallel distributed compensation (PDC) framework. The approach calls for designing feedback gains for each linear time-invariant (LTI) vertex system and then uses the same weightings to form the composite control for the LTI systems:

$$\mathbf{F}(\mathbf{x}(t)) = \mathcal{F} \underset{n=1}{\overset{N}{\boxtimes}} \mathbf{w}_n(x_n(t)), \tag{13.1}$$

155

where $\mathcal{F}$ contains the vertex gains for each LTI system, and $\mathbf{w}_n$ are the corresponding weightings of the system model.

Using linear matrix inequalities (LMIs), various constraints can be prescribed for the closed-loop system:

$$\dot{\mathbf{x}}(t) = (\mathbf{S}(\mathbf{x}(t)) - \mathbf{F}(\mathbf{x}(t)))\,\mathbf{x}(t). \tag{13.2}$$

For LMI-based analysis and design only, the vertex systems and the order of the system (the size of the **A** matrix) are needed to call the command:

```
lmi = lmistruct(S, n);
```

To incorporate multiple controller design constraints into the LMI design process, we can incrementally call the following commands in *TPtool*:

```
lmi = lmi_asym(lmi);
lmi = lmi_asym_decay(lmi, rate);
lmi = lmi_input(lmi, umax, init);
```

meaning that the asymptotic stability constraint is being supplemented with a given decay rate and input constraint.

A TP controller can be designed with `lmi_solve`, which uses efficient convex optimization algorithms internally to find an optimal solution.

```
F = lmi_solve(lmi);
```

Here, F stores the feedback gains for each LTI vertex system. If the LMI optimization has a feasible result, the prescribed stability and performance constraints are guaranteed to hold.

Different convex representations provided by the toolbox can be used with other techniques as well, for example, the command

```
p = topsys(S, n);
```

converts the given TP model S into the polytopic format used by the MATLAB Robust Control Toolbox, which provides many analytical tools for polytopic systems.

Chapter 14

2-D Prototypical Aeroelastic Wing Section with Structural Nonlinearity

This chapter is concerned with designing a state variable feedback controller and observer-based output feedback controller for the prototypical aeroelastic wing section with structural nonlinearity [Bar06b], [Bar06a] using the tensor product (TP) model-based approach proposed in this book. The same control problem extended with friction compensation is given in [TBnt]. The wing section problem has traditionally been used for theoretical as well as experimental analysis of two-dimensional aeroelastic behavior, and shown to exhibit limit-cycle oscillation without control effort. The main advantage of applying the TP approach here is that asymptotic stability via single trailing-edge control surface can be guaranteed. Moreover, in contrast with the previous solutions, a variety of different design specifications, such as decay rate optimization and input constraint, can be readily incorporated. The TP approach further shows that the prototypical aeroelastic wing section can be described exactly by the time-varying convex combination of six linear time-invariant (LTI) models, that is, by a six-element convex polytopic model. Such model representation is the result of the TP model transformation, which has not been investigated in previous studies of areroelastic models. The advantage of this new representation is that it allows application of various recent linear matrix inequality (LMI)-based designs of the wing section for control.

As a final remark, while this chapter utilizes a given two-dimensional analytical prototypical aeroelastic wing section model for design, the methodology can actually be applied to a variety of aeroelastic models of more complex forms. This is because

157

all the steps of the TP model-based design method are executable numerically with reasonable ease, irrelevant of whether the given explicit model is a physical model or just an outcome of a black-box identification. The chapter also presents numerical simulations of the closed-loop performance and compares our results with former alternative control solutions.

14.1   Dynamics modeling

Various studies of aeroelastic systems have appeared in the literature in the past twenty years. For example, a whole series of *Journal of Guidance, Control and Dynamics* has been devoted to detailed studies of this topic. Specifically on the properties of aeroelastic systems, one may refer to the study of free-play nonlinearity by Price et al. in [PLA94] and [PAL95], and by Lee and LeBlanc [LL86]. A complete study of a class of nonlinearities can also be found in [ZY90] and [PAL95]. Further, O'Neil and Strganac [OS95] examined the continuous structural nonlinearity of aeroelastic systems. These works conclude that an aeroelastic system may exhibit a variety of nonlinear phenomena such as limit-cycle oscillation, flutter, and even chaotic vibrations.

For the control of aeroelastic systems, Block and Strganac [BS98] show that in the case of large amplitude limit cycle oscillation behavior, linear control methodologies do not consistently stabilize the aeroelastic systems. At the NASA Langley Research Center, a benchmark active control technique (BACT) wind-tunnel model has been designed, and control algorithms for flutter suspension have been developed by Waszak [Was97], Mukhopadhyay [Muk00], and Joshi and Kelkar [JK00]. Experiments have also been performed on an aeroelastic apparatus in a wind tunnel to examine the effect of nonlinear structural stiffness, and to test control systems that have been designed using linear control theory, feedback linearization technique, and adaptive control strategies [KKS97a], [KSK99], [XS00].

For the two-dimensional prototypical aeroelastic wing section problem, Ko et al. [KKS97b] and Block and Strganac [BS98] have proposed nonlinear feedback control methodologies for a specific class of nonlinear structural effects [OGS96]. In this regard, [KSK99] have also developed a controller via partial-feedback linearization. It has been further shown that by applying an additional control surface, global stabilization can be achieved (see, for example, [KSK98]), and adaptive feedback linearization and the global feedback linearization techniques were introduced for two control actuators in [KSK99], [KKS97a]. Other related works include the Riccati equation-based method in [YSW02] and neural network-based design in [SP00].

In order to be comparable with previous studies of the prototypical aeroelastic wing section, we adopt the same simplified equations of motion as investigated in various works [Fun55], [KKS97a], [XS00], [DCSS78], [KSK98]. We consider the problem of flutter suppression for the prototypical aeroelastic wing section as shown in Figure 14.1.

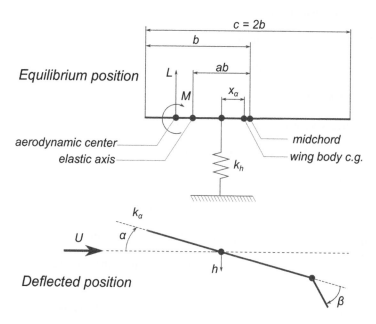

Figure 14.1: Two-dimension flat plate airfoil small deflection, force notation and schematic diagram. (Modified from [Bar06b]–P. Baranyi, "Tensor-product model-based control of two-dimensional aeroelastic system," *Journal of Guidance, Control, and Dynamics*, 29(2):391–400, 2006.)

The variables related to the wing section are defined below:

- h = plunging displacement

- α = pitching displacement

- x_α = the nondimensional distance between elastic axis and the center of mass

- m = the mass of the wing

- I_α = the mass moment of inertia

- b = semichord of the wing

- c_α = the pitch structural damping coefficient

- c_h = the plunge structural damping coefficients

- k_h = the plunge structural spring constant

- $k_\alpha(\alpha)$ = nonlinear stiffness contribution

- L = aerodynamic force

- M = aerodynamic moment

- β = control surface deflection

- ρ = air density

- U = free stream velocity

- c_{l_α} = lift coefficients per angle of attack

- c_{m_α} = moment coefficients per angle of attack

- c_{l_β} = lift coefficients per control surface deflection

- c_{m_β} = moment coefficients per control surface deflection

- a = nondimensional distance from the midchord to the elastic axis

The flat plate airfoil is constrained to have two degrees of freedom (2-DoF), the plunge h and pitch α. The equations of motion can be written as

$$\begin{pmatrix} m & mx_\alpha b \\ mx_\alpha b & I_\alpha \end{pmatrix} \begin{pmatrix} \ddot{h} \\ \ddot{\alpha} \end{pmatrix} + \begin{pmatrix} c_h & 0 \\ 0 & c_\alpha \end{pmatrix} \begin{pmatrix} \dot{h} \\ \dot{\alpha} \end{pmatrix} + \begin{pmatrix} k_h & 0 \\ 0 & k_\alpha(\alpha) \end{pmatrix} \begin{pmatrix} h \\ \alpha \end{pmatrix} = \begin{pmatrix} -L \\ M \end{pmatrix}. \tag{14.1}$$

Here, $k_\alpha(\alpha)$ is obtained by curve fitting on the measured displacement-moment data for nonlinear spring [OS95]:

$$k_\alpha(\alpha) = 2.82(1 - 22.1\alpha + 1315.5\alpha^2 + 8580\alpha^3 + 17289.7\alpha^4).$$

We assume a quasi-steady aerodynamic force and moment as in previous control design approaches:

$$L = \rho U^2 b c_{l_\alpha} \left(\alpha + \frac{\dot{h}}{U} + \left(\frac{1}{2} - a \right) b \frac{\dot{\alpha}}{U} \right) + \rho U^2 b c_{l_\beta} \beta \tag{14.2}$$

$$M = \rho U^2 b^2 c_{m_\alpha} \left(\alpha + \frac{\dot{h}}{U} + \left(\frac{1}{2} - a \right) b \frac{\dot{\alpha}}{U} \right) + \rho U^2 b c_{m_\beta} \beta. \tag{14.3}$$

The above L and M are accurate for the low-velocity regime.

One may refer to the wind tunnel experiments conducted in [BS98]. The system parameters are:

$b = 0.135$ m; $span = 0.6$ m; $k_h = 2844.4$ N/m; $c_h = 27.43$ Ns/m; $c_\alpha = 0.036Ns$;
$\rho = 1.225$ kg/m^3; $c_{l_\alpha} = 6.28$; $c_{l_\beta} = 3.358$; $c_{m_\alpha} = (0.5 + a)c_{l_\alpha}$; $c_{m_\beta} = -0.635$;
$m = 12.387$ kg; $x_\alpha = -0.3533 - a$; $I_\alpha = 0.065$ kg m^2; $c_\alpha = 0.036$;

These data are obtained from experimental models described in full detail in
[KKS97a] and [OS95]. With the flow velocity $u = 15$ m/s and the initial of
$\alpha = 0.1$ rad and $h = 0.01$ m, the resulting time response of the nonlinear system
is depicted in Figure 14.2. Note that the response achieves limit-cycle oscillation as
claimed by [KKS97a], [OS95], [ZY90]. Papers by [OGS96] and [OS95] have also
shown the relations between limit-cycle oscillation, magnitudes, and initial conditions or flow velocities.

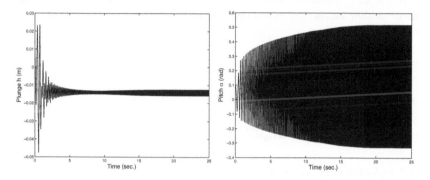

Figure 14.2: Open-loop response for plunge (h) and pitch (α) motion is shown for
$U = 20m/s$ and $a = -0.4$.

Combining Equations (14.1) and (14.2), one obtains:

$$\begin{pmatrix} m & mx_\alpha b \\ mx_\alpha b & I_\alpha \end{pmatrix}\begin{pmatrix} \ddot{h} \\ \ddot{\alpha} \end{pmatrix} + \begin{pmatrix} c_h + \rho U bc_{l_\alpha} & \rho U b^2 c_{l_\alpha}(\frac{1}{2} - a) \\ \rho U b^2 c_{m_\alpha} & c_\alpha - \rho U b^3 c_{m_\alpha}(\frac{1}{2} - a) \end{pmatrix}\begin{pmatrix} \dot{h} \\ \dot{\alpha} \end{pmatrix}$$

$$+ \begin{pmatrix} k_h & \rho U^2 bc_{l_\alpha} \\ 0 & -\rho U^2 b^2 c_{m_\alpha} + k_\alpha(\alpha) \end{pmatrix}\begin{pmatrix} h \\ \alpha \end{pmatrix} = \begin{pmatrix} \rho bc_{l_\beta} \\ \rho b^2 c_{m_\beta} \end{pmatrix}U^2\beta. \qquad (14.4)$$

For control design, the above equation was converted into a state space formulation
in previous studies. Let

$$\mathbf{x}(t) = \begin{pmatrix} x_1(t) \\ x_2(t) \\ x_3(t) \\ x_4(t) \end{pmatrix} = \begin{pmatrix} h \\ \alpha \\ \dot{h} \\ \dot{\alpha} \end{pmatrix} \quad \text{and} \quad \mathbf{u}(t) = \beta.$$

Then we have:

$$\dot{\mathbf{x}}(t) = \mathbf{A}(\mathbf{p}(t))\mathbf{x}(t) + \mathbf{B}(\mathbf{p}(t))\mathbf{u}(t) = \mathbf{S}(\mathbf{p}(t))\begin{pmatrix} \mathbf{x}(t) \\ \mathbf{u}(t) \end{pmatrix}, \qquad (14.5)$$

where

$$\mathbf{A}(\mathbf{p}(t)) = \begin{pmatrix} 0 & 0 & 1 & 0 \\ 0 & 0 & 0 & 1 \\ -k_1 & -(k_2 U^2 + p(x_2(t))) & -c_1(U) & -c_2(U) \\ -k_3 & -(k_4 U^2 + q(x_2(t))) & -c_3(U) & -c_4(U) \end{pmatrix},$$

$$\mathbf{B}(\mathbf{p}(t)) = \begin{pmatrix} 0 \\ 0 \\ g_3 U^2 \\ g_4 U^2 \end{pmatrix},$$

and $\mathbf{p}(t) \in \mathbb{R}^{N=2}$ contains values $x_2(t) = \alpha$ and U. One should note that the equations of motion are also dependent upon the elastic axis location a. The new system variables are

$$d = m(I_\alpha - mx_\alpha^2 b^2); \quad k_1 = \frac{I_\alpha k_h}{d}; \quad k_2 = \frac{I_\alpha \rho b c_{l_\alpha} + mx_\alpha b^3 \rho c_{m_\alpha}}{d}; \quad k_3 = \frac{-mx_\alpha b k_h}{d};$$

$$k_4 = \frac{-mx_\alpha b^2 \rho c_{l_\alpha} - m\rho b^2 c_{m_\alpha}}{d}; \quad p(\alpha) = \frac{-mx_\alpha b}{d} k_\alpha(\alpha); \quad q(\alpha) = \frac{m}{d} k_\alpha(\alpha);$$

$$c_1(U) = \left(I_\alpha(c_h + \rho U b c_{l_\alpha}) + mx_\alpha \rho U^3 c_{m_\alpha}\right)/d;$$

$$c_2(U) = \left(I_\alpha \rho U b^2 c_{l_\alpha}(\tfrac{1}{2} - a) - mx_\alpha b c_\alpha + mx_\alpha \rho U b^4 c_{m_\alpha}(\tfrac{1}{2} - a)\right)/d;$$

$$c_3(U) = \left(-mx_\alpha b c_h - mx_\alpha \rho U b^2 c_{l_\alpha} - m\rho U b^2 c_{m_\alpha}\right)/d;$$

$$c_4(U) = \left(mc_\alpha - mx_\alpha \rho U b^3 c_{l_\alpha}(\tfrac{1}{2} - a) - m\rho U b^3 c_{m_\alpha}(\tfrac{1}{2} - a)\right)/d;$$

$$g_3 = (-I_\alpha \rho b c_{l_\beta} - mx_\alpha b^3 \rho c_{m_\beta})/d; \quad g_4 = (mx_\alpha b^2 \rho c_{l_\beta} + m\rho b^2 c_{m_\beta})/d.$$

14.2   The TP model

We execute the TP model transformation on the quasi-linear parameter-varying (qLPV) state space model (14.5). First, we define the transformation space Ω. We are interested in the interval $U \in [14, 25]m/s$, and we presume that the interval $\alpha \in [-0.1, 0.1]rad$ is also sufficiently large enough. This has practical significance since the prototypical aeroelastic model (14.5) is accurate for low speeds. Therefore, we adopt $\Omega : [14, 25] \times [-0.1, 0.1]$ for the present example. Let the grid density be defined as $M_1 \times M_2$, $M_1 = 101$, and $M_2 = 101$. During execution of the TP model transformation one can see that the rank of the discretized tensor $\mathcal{S}^D \in \mathbb{R}^{M_1 \times M_2 \times 4 \times 4}$ is 3 along the first dimension, and 2 along the second dimension. More specifically, the nonzero singular values in the first dimension are 16808, 1442, and 2, and for the second dimension are 13040 and 7970. Discarding the zero singular values (98 of them in the first dimension and 99 in the second dimension) and keeping all nonzero singular values, the sizes of the resulting matrices $\mathbf{U}_1$ and $\mathbf{U}_2$ are ($M_1 \times 3$) and ($M_2 \times 2$), respectively. In this case, we adopted a tight convex hull using, namely, close-to-normality (CNO) type weighting functions. The weighting functions $w_{1,i}(U)$, $i = 1, \ldots, 3$, and $w_{2,j}(\alpha)$, $j = 1, \ldots, 2$ are depicted in Figure 14.3.

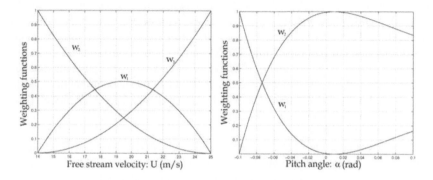

Figure 14.3: Weighting functions of the dimensions U and α.

The corresponding $3 \times 2 = 6$ LTI systems are given as:

$$\mathbf{A}_{1,1} = 10^3 \begin{pmatrix} 0 & 0 & 0.0010 & 0 \\ 0 & 0 & 0 & 0.0010 \\ -0.2314 & -0.0095 & -0.0034 & -0.0001 \\ 0.2780 & -1.1036 & 0.0071 & -0.0000 \end{pmatrix} \quad \mathbf{B}_{1,1} = \begin{pmatrix} 0 \\ 0 \\ -8.5825 \\ -32.4370 \end{pmatrix}$$

$$\mathbf{A}_{2,1} = \begin{pmatrix} 0 & 0 & 1.0000 & 0 \\ 0 & 0 & 0 & 1.0000 \\ -231.3804 & -46.3063 & -4.3776 & -0.2573 \\ 277.9906 & -966.7931 & 10.6520 & 0.4104 \end{pmatrix} \quad \mathbf{B}_{2,1} = \begin{pmatrix} 0 \\ 0 \\ -27.3677 \\ -103.4344 \end{pmatrix}$$

$$\mathbf{A}_{3,1} = 10^3 \begin{pmatrix} 0 & 0 & 0.0010 & 0 \\ 0 & 0 & 0 & 0.0010 \\ -0.2314 & -0.0227 & -0.0039 & -0.0002 \\ 0.2780 & -1.0543 & 0.0089 & 0.0002 \end{pmatrix} \quad \mathbf{B}_{3,1} = 10^3 \begin{pmatrix} 0 \\ 0 \\ -0.0154 \\ -0.0580 \end{pmatrix}$$

$$\mathbf{A}_{1,2} = \begin{pmatrix} 0 & 0 & 1.0000 & 0 \\ 0 & 0 & 0 & 1.0000 \\ -231.3804 & -16.5786 & -3.4333 & -0.1425 \\ 277.9906 & 23.0842 & 7.1447 & -0.0157 \end{pmatrix} \quad \mathbf{B}_{1,2} = \begin{pmatrix} 0 \\ 0 \\ -8.5825 \\ -32.4370 \end{pmatrix}$$

$$\mathbf{A}_{2,2} = \begin{pmatrix} 0 & 0 & 1.0000 & 0 \\ 0 & 0 & 0 & 1.0000 \\ -231.3804 & -53.4094 & -4.3776 & -0.2573 \\ 277.9906 & 159.8695 & 10.6520 & 0.4104 \end{pmatrix} \quad \mathbf{B}_{2,2} = \begin{pmatrix} 0 \\ 0 \\ -27.3677 \\ -103.4344 \end{pmatrix}$$

$$
\mathbf{A}_{3,2} = \begin{pmatrix} 0 & 0 & 1.0000 & 0 \\ 0 & 0 & 0 & 1.0000 \\ -231.3804 & -29.8524 & -3.9054 & -0.1999 \\ 277.9906 & 72.3823 & 8.8983 & 0.1974 \end{pmatrix} \quad \mathbf{B}_{3,2} = \begin{pmatrix} 0 \\ 0 \\ -15.3526 \\ -58.0244 \end{pmatrix}
$$

Note that the rank of the first two dimensions of the discretized tensor $\mathcal{S}^d$ is always 3 and 2 respectively, independently of how we increase the density of the discretization grid. When we numerically check the error between the model (14.5) and the resulting TP model, the error is about 10^{-11}, which is actually caused by round-offs during numerical computation. The TP model as produced thus constitutes an exact description of the original qLPV model (14.5). Moreover, noting that the third singular value of the first dimension is relatively small, one may further decrease the dimensionality by discarding it. This yields a reduced size 2 by 2 TP model, but in this case it is only an approximation to (14.5).

In conclusion, the aeroelastic model (14.5) can be described exactly in finite element convex TP model form with six vertex LTI models. One may actually attempt to derive the weighting functions and the LTI systems analytically from (14.4). The weighting functions of α can be extracted from $k_\alpha(\alpha)$. Finding the weighting functions of U, however, is rather complicated. Finding the tight convex hull via analytic derivations may also be difficult. Regardless, the computation of the TP model transformation takes only a few minutes to generate. It should be emphasized again that the TP model transformation does not depend on the actual analytic formulations of the given model. It can be executed with equal efficiency using a complex analytical physical model, or merely a black-box computational algorithm, of the aeroelastic system. Depending on the circumstances, the resulting TP model may just be an approximation to the aeroelastic system, and contain different and probably a larger number of LTI systems than in the present example. Nevertheless, the TP model as generated will have acceptable accuracy if a sufficient number of nonzero singular values are retained in the process.

14.3   State feedback control design

In the previous section we transformed the aeroelastic model (14.5) into a finite element convex TP model form upon which LMI design under the parallel distributed compensation (PDC) framework can immediately be executed. In this section, we apply the design theorems of Section 11.2 to derive various state feedback controllers, and demonstrate their controlled performance via numerical experiments. In order to be comparable to other published results and figures, the following variables are used in the numerical examples: $a = -0.4$, free stream velocity $U = 20 \; m/s$ (which exceeds the linear flutter velocity $U = 15.5 \; m/s$), and initial values of $h = 0.01 \; m$ and $\alpha = 0.1 \; rad$.

14.3.1 Controller for asymptotic stabilization

Let the resulting LTI vertex systems be substituted into the linear matrix inequalities (LMIs) of Theorem 11.5. The LMIs are feasible in the present case:

$$\mathbf{X} = 10^8 \begin{pmatrix} 0.0213 & 0.0023 & -0.0217 & -0.0506 \\ 0.0023 & 0.0172 & 0.0278 & -0.1131 \\ -0.0217 & 0.0278 & 6.0189 & -0.8293 \\ -0.0506 & -0.1131 & -0.8293 & 4.2114 \end{pmatrix}.$$

The six LTI feedback gains $\mathbf{F}_{i,j}$ are

$$\mathbf{F}_{1,1} = \begin{pmatrix} -6.9079 & 9.8289 & -0.1545 & -1.1896 \end{pmatrix}$$

$$\mathbf{F}_{2,1} = \begin{pmatrix} -2.1000 & 5.7063 & -0.1551 & -0.1848 \end{pmatrix}$$

$$\mathbf{F}_{3,1} = \begin{pmatrix} -3.8807 & 11.2178 & -0.1065 & -0.2071 \end{pmatrix}$$

$$\mathbf{F}_{1,2} = \begin{pmatrix} -5.9511 & -5.1286 & -0.1845 & -0.6240 \end{pmatrix}$$

$$\mathbf{F}_{2,2} = \begin{pmatrix} -2.6178 & -8.2716 & -0.3337 & -0.4773 \end{pmatrix}$$

$$\mathbf{F}_{3,2} = \begin{pmatrix} -4.0982 & -4.1592 & -0.1081 & -0.3394 \end{pmatrix}$$

The control value β (rad) is computed by (11.18) as:

$$u(t) = -\left(\sum_{i=1}^{3} \sum_{j=1}^{2} w_{1,i}(U) w_{2,j}(\alpha) \mathbf{F}_{i,j} \right) \mathbf{x}(t). \tag{14.6}$$

We denote this design as Controller 1. Figure 14.4 shows the corresponding time response of the controlled system. One may observe that asymptotic stabilization is attained and $\alpha(t)$ smoothly converges to zero.

It is important to state that the TP model transformation is a numerical method that can be performed over an arbitrary but bounded domain Ω. Hence, the stability, ensured by the applied LMIs, is restricted to Ω only. This is in line with the fact that the accuracy of a given model is also bounded in reality. In the present example, the prototypical aeroelastic model is accurate for low speeds only, and we have defined Ω accordingly in the design process. The resulting controller guarantees asymptotic stability in $\Omega : [14, 25] \times [-0.1, 0.1]$. If desired, one may extend Ω and execute the design method to check the feasibility again.

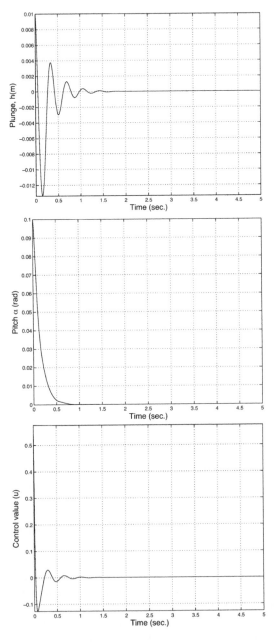

Figure 14.4: Time response of Controller 1 for $U = 20\, m/s$ and $a = -0.4$.

14.3.2 Controller for decay rate control

When we apply Theorem 11.6 in the present case, we find that $\alpha = 0$ (the variable to be maximized in the theorem). This simply means that the decay rate control LMIs in Theorem 11.6 become equivalent to the asymptotic stability LMIs of Theorem 11.5. As a matter of fact, the resulting controller for decay rate control will also be equivalent to Controller 1. Figure 14.4 thus presents also the performance of the decay rate controller.

14.3.3 Controller for constraint on the control value

We augment the LMIs in the above section with those of Theorem 11.7 and solve them altogether to arrive at a design satisfying the constraints on the control value as well. In the case of Controller 2-"min", we search for the minimal bound of control value that still supports a feasible design, that is, the LMIs are feasible. The response of the resulting controller is presented in Figure 14.5. The maximum control value is 0.25. This is significantly smaller than the control value in the case of Controller 1. As a price to pay, we can see that the stabilization time is more than doubled.

For comparison, we also derive the case of Controller 2-"max", where we apply a 10 times larger bound. The response of the resulting controller is presented in Figure 14.6. The maximum control value is 2.15. One can observe that the stabilization time becomes considerably smaller than that of previous cases.

14.3.4 Comparison to other control solutions

We compare our present results with those of the exact feedback linearization technique (see Section IV A and B of [KKS97a]). The reason for picking this technique for comparison is that it applies a single control surface as in our present method. Figure 14.7 presents the closed-loop responses of using the exact feedback linearization controllers. On comparing, one observes that the controllers derived with the TP model and LMI's design in earlier sections of this chapter produce considerably faster responses. One can also observe that α converges smoothly to zero in Figures 14.4, 14.5 and 14.6, unlike the values of α in Figure 14.7. Note that we applied very simple LMI theorems so far. If one would like to go for higher control performance, various choices of performance specifications can be attempted through more powerful LMI design theorems [TW01]. Former solutions of this aeroelastic control problem do not offer such an option to incorporate further control specifications beyond that of stability.

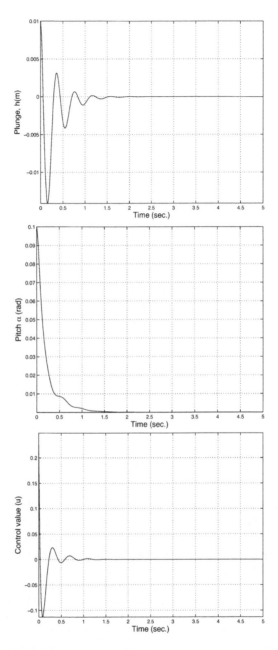

Figure 14.5: Time response of Controller 2-"min" for $U = 20\,m/s$ and $a = -0.4$.

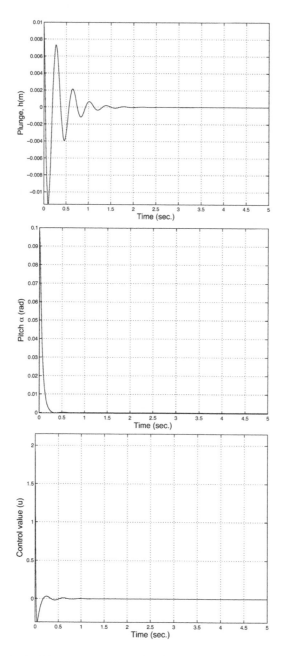

Figure 14.6: Time response of Controller 2-"max" for $U = 20\ m/s$ and $a = -0.4$.

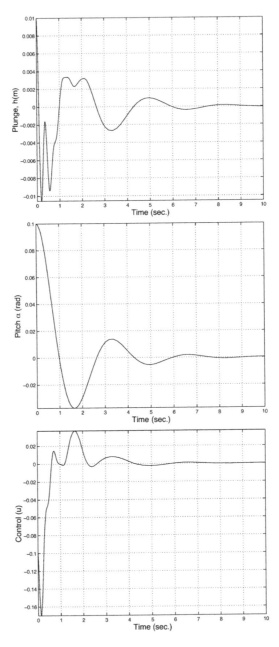

Figure 14.7: Time response by the exact feedback linearization method for $U = 20 \, m/s$ and $a = -0.4$.

14.4 Observer-based output feedback control design

The main objective of this section is to derive an observer for the prototypical aeroelastic wing section via LMI-based design to estimate the unmeasurable state values from the output values, which will then be used in the state feedback controller derived in the previous section. In this regard, we assume that U and the pitch angle $x_2(t)$ are measurable, but that we need to estimate unavailable state values $x_3(t)$ and $x_4(t)$ (see Equation 14.5).

14.4.1 An alternative TP model

This section presents two TP model representations of the prototypical aeroelastic wing section. The first is for the controller design and the second is for the observer design. They will be further used in later sections to demonstrate the effectiveness of convex hull manipulation and to come up with the best pairing of controller and observer designs.

TP MODEL 1: This is the one derived in Section 14.2 adopting CNO type weighting functions (see Figure 14.3).

TP MODEL 2: For observer design, we execute the TP model transformation of (14.5) again by generating inverse normality (INO) and relaxed normality (RNO) (i.e., IRNO) type weighting functions over the same hypergrid and in the same Ω as in the case of TP Model 1. The resulting weighting functions are depicted in Figure 14.8. Here, it is sufficient that we generate IRNO weighting functions in the U-dimension only. The α-dimension weighting functions can still be sum normalization (SN) and nonnegativeness (NN). The resulting $3 \times 2 = 6$ LTI systems are given by

$$
\mathbf{A}_{1,1} = 10^3 \begin{pmatrix} 0 & 0 & 0.0010 & 0 \\ 0 & 0 & 0 & 0.0010 \\ -0.2314 & 0.0127 & -0.0029 & -0.0001 \\ 0.2780 & -1.2050 & 0.0053 & -0.0002 \end{pmatrix} \quad \mathbf{B}_{1,1} = 10^3 \begin{pmatrix} 0 \\ 0 \\ 0.0026 \\ 0.0100 \end{pmatrix},
$$

$$
\mathbf{A}_{2,1} = \begin{pmatrix} 0 & 0 & 1.0000 & 0 \\ 0 & 0 & 0 & 1.0000 \\ -231.3804 & -64.9720 & -4.9178 & -0.3229 \\ 277.9906 & -916.6567 & 12.6582 & 0.6542 \end{pmatrix} \quad \mathbf{B}_{2,1} = \begin{pmatrix} 0 \\ 0 \\ -36.9511 \\ -139.6544 \end{pmatrix},
$$

$$
\mathbf{A}_{3,1} = 10^3 \begin{pmatrix} 0 & 0 & 0.0010 & 0 \\ 0 & 0 & 0 & 0.0010 \\ -0.2314 & -0.0270 & -0.0038 & -0.0002 \\ 0.2780 & -1.0577 & 0.0086 & 0.0002 \end{pmatrix} \quad \mathbf{B}_{3,1} = 10^3 \begin{pmatrix} 0 \\ 0 \\ -0.0176 \\ -0.0664 \end{pmatrix},
$$

$$\mathbf{A}_{1,2} = 10^3 \begin{pmatrix} 0 & 0 & 0.0010 & 0 \\ 0 & 0 & 0 & 0.0010 \\ -0.2314 & -0.0029 & -0.0029 & -0.0001 \\ 0.2780 & 1.2608 & 0.0053 & -0.0002 \end{pmatrix} \quad \mathbf{B}_{1,2} = 10^3 \begin{pmatrix} 0 \\ 0 \\ 0.0026 \\ 0.0100 \end{pmatrix},$$

$$\mathbf{A}_{2,2} = 10^3 \begin{pmatrix} 0 & 0 & 0.0010 & 0 \\ 0 & 0 & 0 & 0.0010 \\ -0.2314 & -0.0805 & -0.0049 & -0.0003 \\ 0.2780 & 1.5491 & 0.0127 & 0.0007 \end{pmatrix} \quad \mathbf{B}_{2,2} = 10^3 \begin{pmatrix} 0 \\ 0 \\ -0.0370 \\ -0.1397 \end{pmatrix},$$

$$\mathbf{A}_{3,2} = 10^3 \begin{pmatrix} 0 & 0 & 0.0010 & 0 \\ 0 & 0 & 0 & 0.0010 \\ -0.2314 & -0.0425 & -0.0038 & -0.0002 \\ 0.2780 & 1.4080 & 0.0086 & 0.0002 \end{pmatrix} \quad \mathbf{B}_{3,2} = 10^3 \begin{pmatrix} 0 \\ 0 \\ -0.0176 \\ -0.0664 \end{pmatrix}.$$

14.4.2   Control system design

This section derives an observer-based output feedback for the prototypical aeroelastic wing section using TP Model 1 for the controller and TP Model 2 for the observer. The observer design is conducted under the PDC framework using the structure discussed in Section 11.2.4. Note that, as formulated, the vector $\mathbf{p}(t)$ does not contain values from the estimated state vector $\hat{\mathbf{x}}(t)$, since $p_1(t)$ equals U and $p_2(t)$ equals the pitch angle ($x_2(t)$), both of which are observable variables. Hence, we estimate only the state values $x_3(t)$ and $x_4(t)$ in our observer. The goal in the present case is thus to determine gains $\mathcal{K}$ in the observer structure (11.14) via Theorem 11.9. We define matrix $\mathbf{C}$ for all r as:

$$\mathbf{y}(t) = \mathbf{C}\mathbf{x}(t),$$

with

$$\mathbf{C}_r = \begin{pmatrix} 1 & 0 & 0 & 0 \\ 0 & 1 & 0 & 0 \end{pmatrix}.$$

Solving the corresponding LMIs then yields the gains $\mathbf{K}_{i,j}$, $i = 1, 2, 3$, $j = 1, 2$, and the state values $x_3(t)$ and $x_4(t)$ are estimated as

$$\dot{\hat{\mathbf{x}}}(t) = \mathbf{A}(\mathbf{p}(t))\hat{\mathbf{x}}(t) + \mathbf{B}(\mathbf{p}(t))u(t) + \left(\sum_{i=1}^{3} \sum_{j=1}^{2} w_{1,i}(U) w_{2,j}(\alpha) \mathbf{K}_{i,j} \right) (\mathbf{y}(t) - \hat{\mathbf{y}}(t)),$$

where

$$\mathbf{y}(t) = \begin{pmatrix} x_1(t) \\ x_2(t) \end{pmatrix} \quad \text{and} \quad \hat{\mathbf{y}}(t) = \begin{pmatrix} \hat{x}_1(t) \\ \hat{x}_2(t) \end{pmatrix} \quad \text{and} \quad \mathbf{p}(t) = \begin{pmatrix} U \\ \alpha \end{pmatrix}.$$

The design process here does not need analytic interaction and it only takes a few minutes on a regular computer. We can easily guarantee other design specifications beyond stability by selecting the proper LMI conditions.

14.4.3 Control performance

Again, simulation for the case $a = -0.4$, $U = 20m/s$, and initial conditions $h = 0.01m$ and $\alpha = 0.1rad$ are conducted in order to be comparable to other published results. The initial observer state is $\hat{\mathbf{x}}(t) = \begin{pmatrix} 0 & 0 & 0 & 0 \end{pmatrix}$. Figure 14.9 shows that the system is asymptotically stabilized. We can see that the stabilization time is a bit longer with the observer than without it, a fact referred to in [Bar06b]. We can also observe that the convergence of the pitch angle is much smoother without the observer in [Bar06b]. If we compare the result derived here to the control result of the feedback linearization (see Figure 7 of [Bar06b]), we can conclude that the output feedback control derived here is faster. Figure 14.10 shows how the system is capable of tracking the trajectory command of the pitch angle. The command trajectory has no practical aspect here; it is only a theoretical command to show the response of the controller system for step and ramp functions. One can see that the pitch angle converges to the command signal in 1 or 2 seconds.

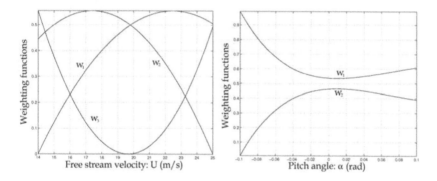

Figure 14.8: Weighting functions of TP Model 2 on the dimensions α and U.

14.5 Convex hull manipulation

This section is devoted to illustrating the process of convex hull manipulation and its effect on LMI-based control system design. We take the IRNO TP Model 2 and CNO TP Model 1 introduced in Section 14.4.1 and perform an interpolation between them. We will then show how the corresponding convex hull will be transformed from an IRNO one in the beginning to a CNO one at the end, and also the effects of such manipulation on control feasibility and performance. The advantage of the TP type polytopic model is that the convex hull can be manipulated in each dimension by the weighting functions. The manipulation is executed in two steps:

Step 1: Generate new weighting functions in between IRNO and CNO types by

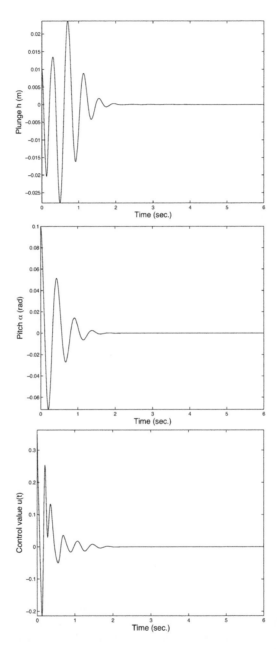

Figure 14.9: Asymptotic stabilization by the derived controller for $U = 20 \, m/s$ and $a = -0.4$.

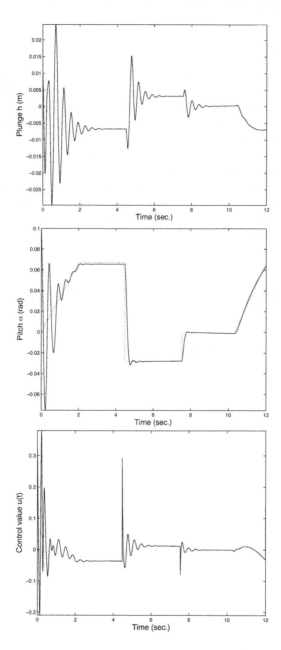

Figure 14.10: Trajectory (dashed line) control of the pitch angle for $U = 20\ m/s$ and $a = -0.4$.

linearly interpolating the weighting functions of TP Model 2 and TP Model 1 (which are IRNO and CNO, respectively):

$$\mathbf{w}_n(\lambda, p_n(t)) = \lambda \cdot \mathbf{w}_n^{CNO}(p_n(t)) + (1 - \lambda) \cdot \mathbf{w}_n^{IRNO}(p_n(t)), \tag{14.7}$$

where λ is a coefficient of value going from 0 to 1. Note that linear interpolation will not affect convexity of the resultant weighting functions. Specifically, we discretize λ in 30 equidistant steps of λ_z between 0 and 1, so we generate $Z = 30$ sets of new weighting functions $\mathbf{w}_{z,n}(p_n(t)) = \mathbf{w}_n(\lambda_z, p_n(t))$ of convexity type going from IRNO and CNO for further investigation.

Step 2: We compute Z sets of LTI vertex systems corresponding to the new weighting functions $\mathbf{w}_z(p_n(t))$ for $z = 1, \dots, Z$. First, we discretize the new weighting functions, then, using the pseudo inverse of these matrices, we calculate the core tensor just like in the case when the convex hull is manipulated in Step 2 of the TP model transformation in Section 3.2.

$$\mathbf{w}_{z,n}(p_n(t)) \Rightarrow \mathcal{S}_z$$

Together, Steps 1 and 2 thus yield $Z = 30$ different TP models embedding IRNO type convexity at $z = 1$ to become CNO type at $z = 30$ but all representing the same system, $\mathbf{S}(\mathbf{p}(t))$:

$$\mathbf{S}(\mathbf{p}(t)) = \mathcal{S}_z \overset{2}{\underset{n=1}{\boxtimes}} \mathbf{w}_{z,n}(p_n(t)). \tag{14.8}$$

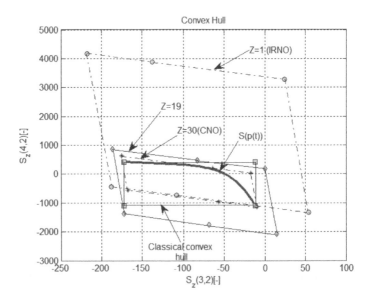

Figure 14.11: Representation of different convex hulls.

14.6   Convex hull geometry

Let us investigate the convex hull of the vertex systems. The number of elements in $S(\mathbf{p}(t))$ is 20, so we need a 20-dimensional space for drawing, which is impossible. To simplify things, we consider only the nonlinear elements of $S(\mathbf{p}(t))$:

$$
S(\mathbf{p}(t)) = \begin{pmatrix} 0 & 0 & 1 & 0 & 0 \\ 0 & 0 & 0 & 1 & 0 \\ -k_1 & S_z(\alpha, U)(3, 2) & S_z(U)(3, 3) & S_z(U)(3, 4) & S_z(U)(3, 5) \\ -k_2 & S_z(\alpha, U)(4, 2) & S_z(U)(4, 3) & S_z(U)(4, 4) & S_z(U)(4, 5) \end{pmatrix}. \quad (14.9)
$$

The number of nonlinear elements is now eight, which is still too high to draw. What we do is thus to present a two-dimensional section of this eight-dimensional space for illustration. To do so, we represent some elements $S_z(3, 2)$ and $S_z(4, 2)$ in a two-dimensional coordinate system. In this coordinate system we draw the elements of $S(\mathbf{p}(t))$ for all possible $\mathbf{p}(t) \in \Omega$, which appear as the bold line in Figure 14.11. Actually this line is a "space" (if we draw the 20 elements it is a 20-dimensional space), where $S(\mathbf{p}(t))$ is varying. The vertices of the IRNO ($z = 1$) type TP model are depicted by dots. The two-dimensional section of the convex hull can be seen by connecting these dots. The two-dimensional section of the CNO ($z = 30$) type convex hull is represented by the connected stars. We can observe that the IRNO type is a wide convex hull, and the CNO type is very tight, significantly tighter than the classical convex hull defined by connected squares.

14.6.1   Effects on LMI-based controller performance

We use output feedback design based on the observer structure (11.14) of Section 14.4.2. We substitute the Z different vertex systems above into Theorem 11.9 to obtain the controller and observer gains simultaneously. We present the effects of convex hulls on the controlled system performance here and the evaluation of the observer separately in the next section. For the cases of $z = 1, \ldots, 18$, the LMIs turn out to be infeasible. Starting from the case of $z = 19$, however, the solution becomes feasible all the way to $z = 30$. This means that only those TP models of $z \geq 19$ yield feasible solutions for the LMIs. Referring to Figure 14.11, the convex hull for $z = 19$ allowing for a feasible LMI design is the one defined by connecting the diamonds.

Figure 14.12 shows the closed-loop performance of the controlled system for $z = 19$, $z = 25$, and $z = 30$. It can be observed that the convex hull upon which the LMI design is based does have certain effects on the closed-loop performance. Particularly, the maximum control torque value for the $z = 30$ (CNO) case is 7 Nm, which is significantly lower than the other two cases.

14.6.2 Effects on LMI-based observer performance

In the previous section we obtained the observer gains simultaneously with the controller gains by solving Theorem 11.9. In order to evaluate the observer we conduct a simulation of the open-loop system response and study how the observer states would approximate the actual system states. Figure 14.13 shows the observer performance to track the open-loop system responses of states h and α for the case $z = 1$, $z = 10$, $z = 20$, and $z = 30$. It can be observed that the best observer performance is attained for $z = 30$, or the CNO type TP model.

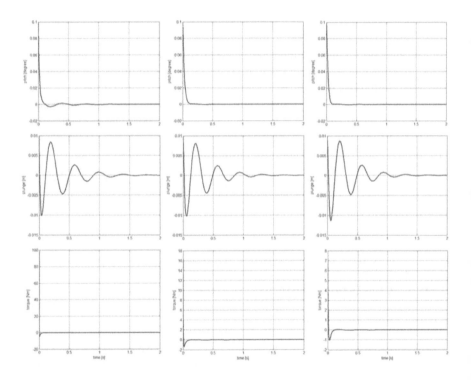

Figure 14.12: Control performance with $U = 20$ m/s and $a = 0.4$ for $z = 19$, $z = 25$, and $z = 30$ (from left to right).

This section demonstrates that the LMIs are very sensitive on the convex hull defined by the convexity of the selected TP model. Upon changing the convex hull, the performance of the controller and also the observer would be changed as well. Manipulation of the convex hull is hence an important consideration with the TP model transformation and LMI-based design process. Note that, in contrast to the present example, in many cases the best control performance is actually achieved when different convex hulls are applied to controller and observer.

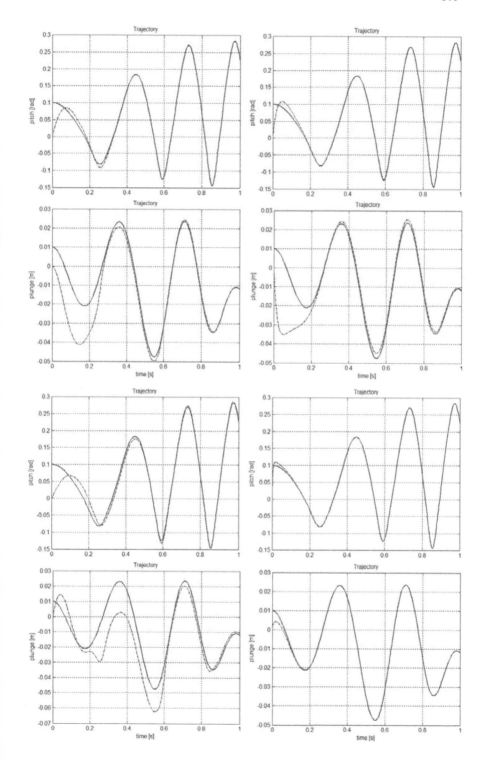

Figure 14.13: Observer performance in open-loop responses, with U = 20 m/s. Top left: z = 1 (IRNO). Top right: z = 10. Bottom left: z = 20. Bottom right z = 30 (CNO). Dotted lines are observer states; continuous lines are actual states.

Chapter 15

3-D Prototypical Aeroelastic Wing Section with Structural Nonlinearity

The aim of this chapter is to show that the steps of the controller design presented in the last chapter for the two degrees of freedom (2-DoF) aeroelastic wing section can be extended in a straightforward manner to the present more realistic and complex three degrees of freedom (3-DoF) aeroelastic wing section. This chapter is based on the work presented in [Tak11].

The topic of controlling a 3-DoF aeroelastic wing section has been well researched. A 3-DoF model of the Nonlinear Aeroelastic Test Apparatus (NATA) with unsteady aerodynamics was first constructed by Block upon which experimental active control results was presented in [BG96] and [BS98]. It was shown that a LQG controller was able to settle limit-cycle oscillations in around 3 seconds.

A robust aeroelastic control of a 3-DoF linear and nonlinear wing-flap system operation in supersonic flight speeds under sliding mode control (SMC) along with the implementation of a sliding mode observer (SMO) was presented in [LCN+10]. Adaptive decoupled fuzzy sliding mode control was designed by Lin and Chin in [LC06]. The approach involves first dividing the aeroelastic system into two subsystems, and then constructing two sliding surfaces using the state variables of the decoupled system. An intermediate variable is introduced to incorporate these two sliding surfaces. Gujjula and Singh in [SG05] deal with variable structure control of unsteady aeroelastic systems with partial state information.

Prime et al. in [PCD08] presented a mixed H_∞/H_2 scheduling control scheme for a 3-DoF aeroelastic system under varying airspeed and gust conditions. In their work, the dynamics are transformed into a linear fractional representation (LFR) such that the nonlinear effects of airspeed on the dynamics act as a gain feedback to the nominal system. A controller in LFR, which allows it to schedule with airspeed, is synthesized using linear matrix inequalities (LMIs). Linear parameter-varying (LPV) control of an improved 3-DoF aeroelastic model is further discussed in [PCDS10].

15.1   Dynamics modeling

We adopt the same equations of motion as in the works of [PCD08], [PCDS10] so as to facilitate ready comparison with previous studies of the prototypical aeroelastic wing section. The new variables, which differ from the variables listed for the case of the 2-DoF aeroelastic wing section, are defined below:

- m_h mass of the translating components of the carriage

- m_α mass of the wing and the rotational component of the carriage

- m_β mass of the trailing-edge

- $c_{\beta_{servo}}$ virtual damping provided by the trailing-edge servo motor

- $k_{\beta_{servo}}$ virtual stiffness provided by the trailing-edge servo motor

- β_{des} trailing-edge control signal

- $\hat{I}_\alpha$ replaced rotational moment of inertia of the wing and rotational component of the carriage

- $\hat{I}_\beta$ replaced rotational moment of inertia of the trailing-edge section

- r_β distance between the trailing-edge pivot and center of gravity

The flat plate airfoil is constrained to have three DoF, which are the plunge h, pitch α, and trailing-edge surface deflection β. The equations of motion can be written as

$$
\begin{pmatrix}
m_h + m_\alpha + m_\beta & m_\alpha x_\alpha b + m_\beta r_\beta + m_\beta x_\beta & m_\beta r_\beta \\
m_\alpha x_\alpha b + m_\beta r_\beta + m_\beta x_\beta & \hat{I}_\alpha + \hat{I}_\beta + m_\beta r_\beta^2 + 2x_\beta m_\beta r_\beta & \hat{I}_\beta + x_\beta m_\beta r_\beta \\
m_\beta r_\beta & \hat{I}_\beta + x_\beta m_\beta r_\beta \hat{I}_\beta m x_\alpha b & \hat{I}_\alpha
\end{pmatrix}
\begin{pmatrix} \ddot{h} \\ \ddot{\alpha} \\ \ddot{\beta} \end{pmatrix} + \quad (15.1)
$$

$$
\begin{pmatrix}
c_h & 0 & 0 \\
0 & c_\alpha & 0 \\
0 & 0 & c_{\beta_{servo}}
\end{pmatrix}
\begin{pmatrix} \dot{h} \\ \dot{\alpha} \\ \dot{\beta} \end{pmatrix}
+
\begin{pmatrix}
k_h & 0 & 0 \\
0 & k_\alpha(\alpha) & 0 \\
0 & 0 & k_{\beta_{servo}}
\end{pmatrix}
\begin{pmatrix} h \\ \alpha \\ \beta \end{pmatrix}
=
\begin{pmatrix} -L \\ M \\ k_{\beta_{servo}}\beta_{des} \end{pmatrix}.
$$

Here, $k_\alpha(\alpha)$ is obtained by curve fitting the measured displacement-moment data for nonlinear spring [PCDS10]:

$$k_\alpha(\alpha) = 25.55 - 103.19\alpha + 543.24\alpha^2.$$

We assume the quasi-steady aerodynamic force and moment as in previous control design approaches:

$$L = \rho U^2 b C_{l_\alpha}\left(\alpha + \frac{\dot h}{U} + \left(\frac{1}{2} - a\right) b \frac{\dot\alpha}{U}\right) + \rho U^2 b c_{l_\beta}\beta \tag{15.2}$$

$$M = \rho U^2 b^2 C_{m_{\alpha,eff.}}\left(\alpha + \frac{\dot h}{U} + \left(\frac{1}{2} - a\right) b \frac{\dot\alpha}{U}\right) + \rho U^2 b C_{m_{\beta,eff}}\beta.$$

The above L and M are accurate for the low-velocity regime.

Based on [PCDS10], we assume that the trailing-edge servo motor dynamics can be represented by a second-order system of the form:

$$\hat I_\beta \ddot\beta + c_{\beta_{servo}}\dot\beta + k_{\beta_{servo}}\beta = k_{\beta_{servo}}u_\beta. \tag{15.3}$$

Combining Equations (15.1), (15.3), and (15.2) one obtains:

$$\underbrace{\begin{pmatrix} m_h + m_\alpha + m_\beta & m_\alpha x_\alpha b + m_\beta r_\beta + m_\beta x_\beta & m_\beta r_\beta \\ m_\alpha x_\alpha b + m_\beta r_\beta + m_\beta x_\beta & \hat I_\alpha + \hat I_\beta + m_\beta r_\beta^2 + 2x_\beta m_\beta r_\beta & \hat I_\beta + x_\beta m_\beta r_\beta \\ m_\beta r_\beta & \hat I_\beta + x_\beta m_\beta r_\beta \hat I_\beta m x_\alpha b & I_\alpha \end{pmatrix}}_{\mathbf{M}_{eom}} \begin{pmatrix} \ddot h \\ \ddot\alpha \\ \ddot\beta \end{pmatrix} + \tag{15.4}$$

$$\underbrace{\begin{pmatrix} c_h + \rho b S C_{l_\alpha} U & \left(\frac{1}{2} - a\right) b \rho b S C_{l_\alpha} U & 0 \\ -\rho b^2 S C_{m_{\alpha,eff}} U & c_\alpha - \left(\frac{1}{2} - a\right) b \rho b^2 S C_{m_{\alpha,eff}} U & 0 \\ 0 & 0 & c_{\beta_{servo}} \end{pmatrix}}_{\mathbf{C}_{eom}} \begin{pmatrix} \dot h \\ \dot\alpha \\ \dot\beta \end{pmatrix} + $$

$$\underbrace{\begin{pmatrix} k_h & \rho b S C_{l_\alpha} U^2 & \rho b S C_{l_\beta} U^2 \\ 0 & k_\alpha(\alpha) - \rho b^2 S C_{m_{\alpha,eff}} U^2 & -\rho b^2 S C_{m_{\beta,eff}} U^2 \\ 0 & 0 & k_{\beta_{servo}} \end{pmatrix}}_{\mathbf{K}_{eom}} \begin{pmatrix} h \\ \alpha \\ \beta \end{pmatrix} = \underbrace{\begin{pmatrix} 0 \\ 0 \\ k_{\beta_{servo}} \end{pmatrix}}_{\mathbf{F}_{eom}} \mathbf{u}, \tag{15.5}$$

where $\mathbf{M}_{eom}$ is the mass matrix of the equation of motion, $\mathbf{C}_{eom}$ is the damping matrix of the equation of motion, $\mathbf{K}_{eom}$ is the stiffness matrix of the equation of motion, and $\mathbf{F}_{eom}$ is the forcing matrix of the equation of motion.

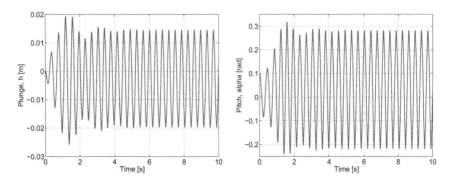

Figure 15.1: Open-loop response for plunge (h) and pitch (α) motion at $U = 14.1 m/s$.

For control design, the above equation was converted into the state space formulation in previous studies. Let

$$\mathbf{x}(t) = \begin{pmatrix} x_1(t) \\ x_2(t) \\ x_3(t) \\ x_4(t) \\ x_5(t) \\ x_6(t) \end{pmatrix} = \begin{pmatrix} \dot{h} \\ \dot{\alpha} \\ \dot{\beta} \\ h \\ \alpha \\ \beta \end{pmatrix} \quad \text{and} \quad \mathbf{u}(t) = u_\beta.$$

The state matrix is then

$$\mathbf{A}(\mathbf{p}(t)) = \begin{pmatrix} -\mathbf{M}_{eom}^{-1}\mathbf{C}_{eom}(\mathbf{p}(t)) & -\mathbf{M}_{eom}^{-1}\mathbf{K}_{eom}(\mathbf{p}(t)) \\ -\mathbf{I} & 0 \end{pmatrix}. \tag{15.6}$$

The input matrix is

$$\mathbf{B} = \begin{pmatrix} \mathbf{M}_{eom}^{-1}\mathbf{F}_{eom} \\ 0 \end{pmatrix}. \tag{15.7}$$

We assume that only state variable $x_5(t) = \alpha$ is measurable, thus we choose the output to be $\mathbf{y} = x_5(t) = \alpha$. Actually, α is one of the two variables in $\mathbf{p}(t) \in \mathbb{R}^2$ together with U. The output matrix is the following:

$$\mathbf{C} = \begin{pmatrix} 0 & 0 & 0 & 0 & 1 & 0 \end{pmatrix}. \tag{15.8}$$

The feed-through matrix is zero:
$$\mathbf{D} = 0. \tag{15.9}$$

The system matrix can be constructed in the following way:

$$\mathbf{S}(\mathbf{p}(t)) = \begin{pmatrix} \mathbf{A}(\mathbf{p}(t)) & \mathbf{B} \\ \mathbf{C} & \mathbf{D} \end{pmatrix}. \tag{15.10}$$

The system parameters we used for simulations can be found in [PCDS10] and they are listed in the following:

$m_h = 6.516\ kg$; $m_\alpha = 6.7\ kg$; $m_\beta = 0.537\ kg$; $x_\alpha = 0.21$; $x_\beta = 0.233$; $r_\beta = 0\ m$; $a = -0.673\ m$; $b = 0.1905\ m$; $\hat{I}_\alpha = 0.126\ kgm^2$; $\hat{I}_\beta = 10^{-5}$; $c_h = 27.43\ Nms/rad$; $c_\alpha = 0.215\ Nms/rad$; $c_{\beta_{servo}} = 4.182 * 10^{-4}\ Nms/rad$; $k_h = 2844$; $k_{\beta_{servo}} = 7.6608 *$ 10^{-3}; $\rho = 1.225\ kg/m^3$; $C_{l_\alpha} = 6.757$; $C_{m_{\alpha,eff}} = -1.17$; $C_{l_\beta} = 3.774$; $C_{m_{\beta,eff}} = -2.1$; $S = 0.5945\ m$;

Figure 15.1 shows the open-loop response of the developed model at free stream velocity of $U = 14.1 m/s$.

15.2   The TP model

The tensor product (TP) model for the 3-DoF prototypical aeroelastic wing section is obtained based on the same procedures as for the 2-DoF version in the last chapter by applying the TP model transformation on the quasi-LPV (qLPV) state space model (15.10). We first define the transformation space Ω. We set the interval $U \in [8, 20](m/s)$ and presume that the interval $\alpha \in [-0.3, 0.3](rad)$ is sufficiently large enough. This has practical significance since the prototypical aeroelastic model (15.10) is accurate for low speeds. Therefore, let $\Omega : [-0.3, 0.3] \times [8, 20]$ in the present example. Moreover, we let the grid density be $M_1 \times M_2$, with $M_1 = 137$ and $M_2 = 137$. During execution of the TP model transformation one can see that the rank of the discretized tensor $S^D \in \mathbb{R}^{M_1 \times M_2 \times 6 \times 6}$ on the first dimension is 2 and on the second dimension is 3. The nonzero singular values of the first dimension are 192064 and 29878.8. The nonzero singular values of the second dimension are 194303, 5254.15, and 4.16. Discarding only zero singular values (we discard 135 zero singular values on the first and 134 zero singular values on the second dimensions), and keeping all nonzero singular values, the sizes of the resulting matrices $\mathbf{U}_1$ and $\mathbf{U}_2$ are $(M_1 \times 2)$ and $(M_2 \times 3)$, respectively. The weighting functions $w_{1,i}(\alpha)$, $i = 1, 2$, and $w_{2,j}(U)$, $j = 1, 2, 3$, are depicted in Figure 15.2. In this regard, we choose close-to-normality (CNO) type weighting functions for a tight convex hull. We can observe that the weighting functions are basically the same as those of the 2-DoF aeroelastic wing section in Figure 14.3.

The $2 \times 3 = 6$ linear time-invariant (LTI) vertex systems are given by:

$$\mathbf{A}_{1,1} = 10^2 \begin{pmatrix} -0.0309 & 0.0003 & -0.0000 & -2.2294 & 0.0953 & -0.0508 \\ 0.0504 & -0.0183 & 0.0000 & 5.6492 & -7.3719 & -0.4764 \\ -0.0504 & 0.0115 & -0.4182 & -5.6492 & 7.3719 & -7.1843 \\ 0.0100 & 0 & 0 & 0 & 0 & 0 \\ 0 & 0.0100 & 0 & 0 & 0 & 0 \\ 0 & 0 & 0.0100 & 0 & 0 & 0 \end{pmatrix},$$

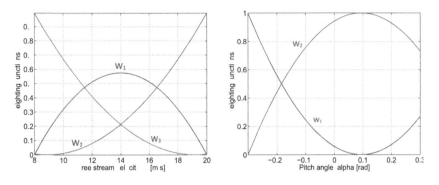

Figure 15.2: CNO type weighting functions over U and α for the generated TP model.

$$\mathbf{A}_{2,1} = 10^2 \begin{pmatrix} -0.0309 & 0.0003 & -0.0000 & -2.2294 & -0.0730 & -0.0508 \\ 0.0504 & -0.0183 & 0.0000 & 5.6492 & -1.4835 & -0.4764 \\ -0.0504 & 0.0115 & -0.4182 & -5.6492 & 1.4835 & -7.1843 \\ 0.0100 & 0 & 0 & 0 & 0 & 0 \\ 0 & 0.0100 & 0 & 0 & 0 & 0 \\ 0 & 0 & 0.0100 & 0 & 0 & 0 \end{pmatrix},$$

$$\mathbf{A}_{1,2} = 10^2 \begin{pmatrix} -0.0350 & 0.0003 & -0.0000 & -2.2294 & -0.0612 & -0.1206 \\ 0.0487 & -0.0229 & 0.0000 & 5.6492 & -7.4386 & -1.1314 \\ -0.0487 & 0.0069 & -0.4182 & -5.6492 & 7.4386 & -6.5293 \\ 0.0100 & 0 & 0 & 0 & 0 & 0 \\ 0 & 0.0100 & 0 & 0 & 0 & 0 \\ 0 & 0 & 0.0100 & 0 & 0 & 0 \end{pmatrix},$$

$$\mathbf{A}_{2,2} = 10^2 \begin{pmatrix} -0.0350 & 0.0003 & -0.0000 & -2.2294 & -0.2296 & -0.1206 \\ 0.0487 & -0.0229 & 0.0000 & 5.6492 & -1.5501 & -1.1314 \\ -0.0487 & 0.0069 & -0.4182 & -5.6492 & 1.5501 & -6.5293 \\ 0.0100 & 0 & 0 & 0 & 0 & 0 \\ 0 & 0.0100 & 0 & 0 & 0 & 0 \\ 0 & 0 & 0.0100 & 0 & 0 & 0 \end{pmatrix},$$

$$\mathbf{A}_{1,3} = 10^2 \begin{pmatrix} -0.0268 & 0.0004 & -0.0000 & -2.2294 & 0.1675 & -0.0186 \\ 0.0522 & -0.0161 & 0.0000 & 5.6492 & -7.3412 & -0.1743 \\ -0.0522 & 0.0136 & -0.4182 & -5.6492 & 7.3412 & -7.4864 \\ 0.0100 & 0 & 0 & 0 & 0 & 0 \\ 0 & 0.0100 & 0 & 0 & 0 & 0 \\ 0 & 0 & 0.0100 & 0 & 0 & 0 \end{pmatrix},$$

$$\mathbf{A}_{2,3} = 10^2 \begin{pmatrix} -0.0268 & 0.0004 & -0.0000 & -2.2294 & -0.0007 & -0.0186 \\ 0.0522 & -0.0161 & 0.0000 & 5.6492 & -1.4527 & -0.1743 \\ -0.0522 & 0.0136 & -0.4182 & -5.6492 & 1.4527 & -7.4864 \\ 0.0100 & 0 & 0 & 0 & 0 & 0 \\ 0 & 0.0100 & 0 & 0 & 0 & 0 \\ 0 & 0 & 0.0100 & 0 & 0 & 0 \end{pmatrix},$$

and

$$\mathbf{B}_{1,1} = \mathbf{B}_{2,1} = \mathbf{B}_{1,2} = \mathbf{B}_{2,2} = \mathbf{B}_{1,3} = \mathbf{B}_{2,3} = 10^2 \begin{pmatrix} 0.0000 \\ -0.0005 \\ 7.6613 \\ 0 \\ 0 \\ 0 \end{pmatrix}.$$

Again, as for the 2-DoF case, the rank of the first two dimensions of the discretized tensor $\mathcal{S}^d$ is always 2 and 3, respectively, regardless of the density of the discretization grid. Numerically checking the error between the model (15.10) and the resulting TP model yields an error of roughly 10^{-11}, which is mainly caused by numerical computation. The TP model here is hence an exact description of the original model (15.10). The third singular value of the second dimension is very small relatively. Discarding it yields a reduced dimensionality system having two weighting functions for each dimension, constituting an approximation to the model (15.10).

In conclusion, the aeroelastic model (15.10) can be described exactly in finite element convex TP model form with six vertex LTI models. Again, one may try to derive the weighting functions and the LTI systems analytically from (15.5) but it will be a very tedious process.

15.3 LMI-based output feedback control design

In the previous chapter we transformed the 2-DoF aeroelastic model (14.5) into finite element convex TP model form, whereupon linear matrix inequality (LMI) design theorems under the parallel distributed compensation (PDC) framework in Section 11.2 can be readily executed. In this section, we repeat the same process to the 3-DoF version to derive an output feedback controller. For comparison with other existing works, the numerical examples are performed with free stream velocity $U = 14.1m/s$. In the simulation, the controller is off until after the oscillations are fully developed at $t = 10s$. The following figures show only the part of the simulation where the controller is turned on.

Upon the feasibility test of the selected LMIs, the design theorems yield the control as

$$u(t) = -\mathcal{F} \underset{n=1}{\overset{N}{\boxtimes}} \mathbf{w}_n(p_n(t))\mathbf{x}(t) \tag{15.11}$$

where $\mathcal{F}$ contains the LTI vertex feedback gain corresponding to each of the LTI vertex systems of the TP model.

At this juncture, we desire an observer to estimate the unmeasurable states from the output values to be used with the above control (15.11). We assume that U and the pitch angle ($x_5(t)$) are measurable, but we need to estimate unavailable state values $x_1(t), x_2(t), x_3(t), x_4(t)$, and $x_6(t)$ from them.

The observer and the controller are derived here as for the 2-DoF wing section using Theorem 11.9 and the TP model in Section 15.2. At this point, we should emphasize that in the present study, as required, the vector $\mathbf{p}(t)$ does not contain values from the estimated state vector $\hat{\mathbf{x}}(t)$, since $p_1(t) = \alpha$ equals $x_5(t)$ and $p_2(t) = U$, and are observable here. Consequently, the goal in the present case is to determine gains $\mathcal{K}$ in the observer structure (11.14). For each of the six vertex systems, we define matrix $\mathbf{C}_{i,j}$ for all $i = 1, 2$ and $j = 1, 2, 3$ as

$$\mathbf{C}_{i,j} = \begin{pmatrix} 0 & 0 & 0 & 0 & 1 & 0 \end{pmatrix}$$

and

$$\mathbf{y}(t) = \mathbf{C}_{\mathbf{i},\mathbf{j}}\mathbf{x}(t).$$

The observer gains and the controller gains are to be solved simultaneously using Theorem 11.9 and related ones with additional LMIs as needed. If successful, we obtain the gains $\mathbf{K}_{i,j}$ to estimate the state values as follows:

$$\hat{\mathbf{x}}(t) = \mathbf{A}(\mathbf{p}(t))\hat{\mathbf{x}}(t) + \mathbf{B}(\mathbf{p}(t))u(t) + \left(\sum_{i=1}^{2}\sum_{j=1}^{3}w_{1,i}(\alpha)w_{2,j}(U)\mathbf{K}_{i,j}\right)(\mathbf{y}(t) - \hat{\mathbf{y}}(t)),$$

where

$$\mathbf{y}(t) = \begin{pmatrix} x_5(t) \end{pmatrix} \quad \text{and} \quad \hat{\mathbf{y}}(t)' = \begin{pmatrix} \hat{x}_1(t) & \hat{x}_2(t) & \hat{x}_3(t) & \hat{x}_4(t) & \hat{x}_6(t) \end{pmatrix} \quad \text{and} \quad \mathbf{p}(t) = \begin{pmatrix} \alpha \\ U \end{pmatrix}.$$

The $\hat{\mathbf{x}}(t)$ will be used in the place of $\mathbf{x}$ in (15.11) for various types of control described below.

15.3.1   Controller 1: Asymptotic stabilization

The feasibility test of Theorem 11.9 leads to the following controller and observer designs.

The six LTI feedback gains $\mathbf{F}_{i,j}$ are

$$\mathbf{F}_{1,1} = 10^3 \left(-0.0591 \quad -0.3445 \quad 0.0025 \quad -5.9181 \quad -2.6278 \quad 0.3944\right),$$

$$\mathbf{F}_{2,1} = 10^3 \left(-0.0527 \quad -0.3026 \quad 0.0022 \quad -5.2087 \quad -2.3039 \quad 0.3475\right),$$

$$\mathbf{F}_{1,2} = 10^3 \left(-0.0305 \quad -0.1742 \quad 0.0012 \quad -3.0258 \quad -1.3152 \quad 0.2037\right),$$

$$\mathbf{F}_{2,2} = 10^3 \left(-0.0369 \quad -0.2161 \quad 0.0016 \quad -3.7352 \quad -1.6391 \quad 0.2506\right),$$

$$\mathbf{F}_{1,3} = 10^3 \left(-0.0629 \quad -0.3618 \quad 0.0026 \quad -6.2150 \quad -2.7597 \quad 0.4139\right),$$

$$\mathbf{F}_{2,3} = 10^3 \left(-0.0693 \quad -0.4037 \quad 0.0030 \quad -6.9244 \quad -3.0836 \quad 0.4608\right).$$

The six LTI observer gains $\mathbf{K}_{i,j}$ are

$$\mathbf{K}_{1,1} = 10^3 \left(0.7868 \quad 8.9319 \quad -2.2086 \quad 0.0796 \quad 0.2391 \quad -0.0966\right),$$

$$\mathbf{K}_{2,1} = 10^3 \left(0.7700 \quad 9.5208 \quad -2.7975 \quad 0.0796 \quad 0.2391 \quad -0.0966\right),$$

$$\mathbf{K}_{1,2} = 10^3 \left(0.7578 \quad 8.8487 \quad -2.1776 \quad 0.0784 \quad 0.2365 \quad -0.0924\right),$$

$$\mathbf{K}_{2,2} = 10^3 \left(0.7410 \quad 9.4376 \quad -2.7665 \quad 0.0784 \quad 0.2365 \quad -0.0924\right),$$

$$\mathbf{K}_{1,3} = 10^3 \left(0.7996 \quad 8.9682 \quad -2.2222 \quad 0.0801 \quad 0.2403 \quad -0.0985\right),$$

$$\mathbf{K}_{2,3} = 10^3 \left(0.7827 \quad 9.5571 \quad -2.8110 \quad 0.0801 \quad 0.2403 \quad -0.0985\right).$$

The control value $u(t)$ (rad) is computed from (15.11) using $\hat{\mathbf{x}}(t)$ as

$$u(t) = -\left(\sum_{i=1}^{2}\sum_{j=1}^{3} w_{1,i}(U)w_{2,j}(\alpha)\mathbf{F}_{i,j}\right)\hat{\mathbf{x}}(t). \tag{15.12}$$

The performance of the observer was tested by simulation. The results are given in Figure 15.3 for the initial states of $\mathbf{x}(t) = [0, 0, 0, 0, 0, 1, 0]$ and $\hat{\mathbf{x}}(t) = [0, 0, 0, 0, 0, 0]$. Figure 15.4 shows the time response of the controlled system.

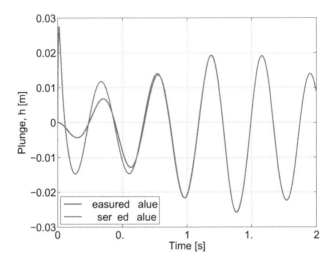

Figure 15.3: Observer test.

15.3.2   Controller 2: Constraint on the control value

We derive two additional controllers as in the previous chapter. In order to bound the control values we add Theorem 11.7 to the design. In the case of Controller 2-"min" we searched the minimal bound of the control value while keeping the LMIs feasible. The upper bound of the initial condition is set at 2. Figure 15.5 shows the response of the resulting controller.

For comparison we derive Controller 2-"max", where we increase the bound drastically like in the 2-DoF case. The response of the resulting controller is presented in Figure 15.6. One can observe that the control signal becomes considerable larger than that in the previous figure.

15.3.3   Control performance

The control performance here is similar to the cases of the 2-DoF aeroelastic wing section. One can observe the asymptotic stabilization of the system and that α smoothly converges to zero. An important issue should be addressed here. The applied LMIs guarantee that the resulting controller is stable. However, the TP model transformation is a numerical method that can be performed over an arbitrary, but bounded domain Ω. Therefore, the stability ensured by the applied LMIs is restricted to Ω. Note that the accuracy of the given model is also bounded in reality. In the present example the prototypical aeroelastic model is accurate for low speeds only,

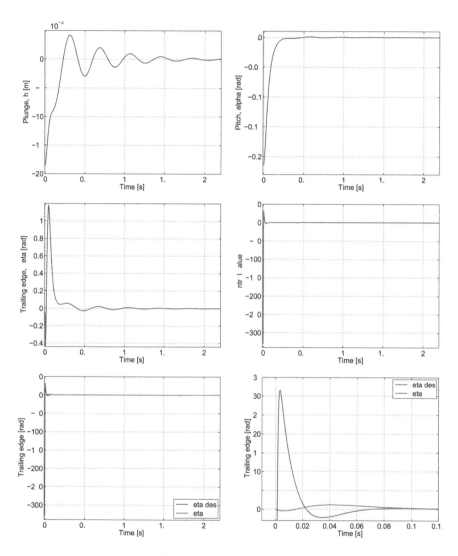

Figure 15.4: Time response of Controller 1 for $U = 14.4m/s$.

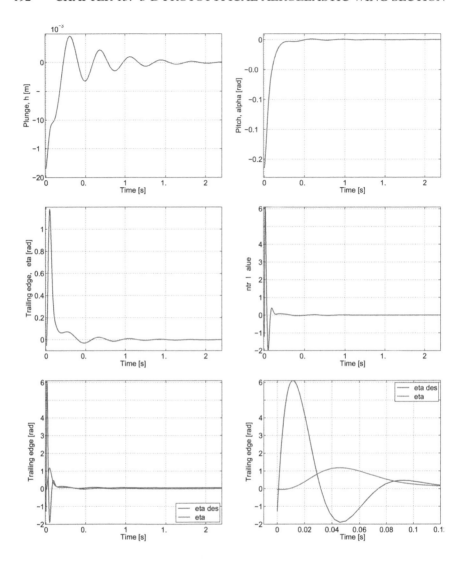

Figure 15.5: Time response of Controller 2-"min" for $U = 14.4m/s$.

and we have defined Ω accordingly in the design process. The resulting controller guarantees asymptotic stability in $\Omega : [-0.3, 0.3] \times [8, 20]$. One may extend Ω and execute the design method again.

The control signal in the case of Controller 1 is rather high for real application. However, Controller 2-"min" achieves the stabilization of the wing with a maximal control value of six. Increasing the bound of the control signal by 10 times leads to Controller 2-"max". Stabilizing the aeroelastic wing section with Controller 2-

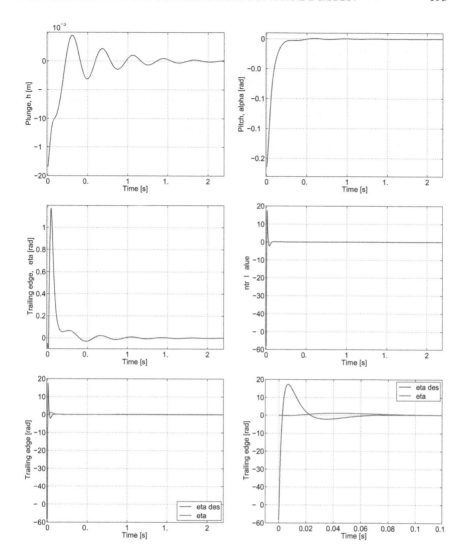

Figure 15.6: Time response of Controller 2-"max" for $U = 14.4m/s$.

"max" implies approximately 60 as the maximal control value, which is also 10 times higher than the maximal control value of Controller 2-"min". We can conclude that Controller 2-"min" out of the three designed controllers has the best performance.

We can compare the control performance with the results presented in [PCDS10], where the LQR controller was designed for the qLPV model of the 3-DoF aeroelastic wing section. On comparing, one observes that the controllers derived with the TP model and LMI design in earlier sections of this chapter produce considerably faster responses, but the cost is a higher control value. We also have to mention that the qLPV model in [PCDS10] has nonlinearity only in one dimension, namely in U.

Note that we have applied very simple LMI theorems so far. If one would like to go for higher control performance, various choices of performance specifications can be attempted through more powerful LMI design theorems. Former solutions of this aeroelastic control problem do not offer such an option to incorporate further control specifications beyond that of stability.

Chapter 16

3-DoF Helicopter with Four Propellers

In this chapter, we apply the tensor product (TP) model-based approach to derive a controller to stabilize a three degrees of freedom (3-DoF) remote controlled (RC) helicopter. The nonlinear model of the RC helicopter, and its simplified version to facilitate the control design, are presented first. Then, the TP model transformation is executed on the simplified model, upon which the parallel distributed compensation (PDC) framework can be applied. Specifically, we consider the incorporation of three practical control specifications via properly selected linear matrix inequalities (LMIs). They include decay rate control to achieve good speed of response, constraint over the control input of the propellers to avoid actuator saturations, and the design of a robust controller to cover model discrepancies due to model simplification. Numerical studies showing that the resulting controller satisfies the desired control specifications are also conducted. We note that the designed controller here is equivalent to the one derived in [TOW04] that was successfully tested in laboratory experiments. This chapter is based on the work of [BKT09].

16.1 Dynamics Modeling

Figure 16.1 shows the picture of a 3-DoF RC helicopter. In order to focus on the stabilization issue (and to possibly carry out laboratory testing), the helicopter is fixed at a (joint) point shown in the top left subfigure of Figure 16.2, hereby reducing some of its DoFs. Other subfigures in Figure 16.2 present the helicopter from different perspectives to help indicate the pitch, roll, and yaw modes of the vehicle and their corresponding physical parameters.

The equations of motion for the RC helicopter fixed at a joint point are

$$Me^2\ddot{\gamma}(t) + I_\gamma\ddot{\gamma}(t) - Mge\sin\gamma(t)$$
$$= F_1(t)\sqrt{l_1^2 + e^2}\cos\left(\tan^{-1}\frac{e}{l_1}\right) - F_3(t)\sqrt{l_1^2 + e^2}\cos\left(\tan^{-1}\frac{e}{l_1}\right), \quad (16.1)$$

$$Me^2\ddot{\beta}(t) + I_\beta\ddot{\beta}(t) - Mge\sin\beta(t)$$
$$= -F_2(t)\sqrt{l_2^2 + e^2}\cos\left(\tan^{-1}\frac{e}{l_2}\right) + F_4(t)\sqrt{l_2^2 + e^2}\cos\left(\tan^{-1}\frac{e}{l_2}\right), \quad (16.2)$$

$$(I_\alpha + 4I_1)\ddot{\alpha}(t) = I_1\left(\ddot{\theta}_1(t) - \ddot{\theta}_2(t) + \ddot{\theta}_3(t) - \ddot{\theta}_4(t)\right), \quad (16.3)$$

where $\gamma(t)$, $\beta(t)$, and $\alpha(t)$ are the roll angle, pitch angle, and yaw angle, respectively. M is the mass of the helicopter. I_γ, I_β, and I_α are the moments of inertia around x, y, and z axes with respect to the gravity point of the helicopter, respectively. I_1 is the moment of inertia of a propeller, g is the gravity constant, and e, l_1, and l_2 are the lengths shown in Figure 16.2. F_i is the lift force generated by the ith propeller described by

$$F_i(t) = \frac{1}{3}C_L\rho S l_w^2\dot{\theta}_i^2(t), \quad (16.4)$$

where C_L is the lift coefficient, ρ is the air density, and S and l_w are the area and length, respectively, of a wing of each propeller. $\dot{\theta}_i(t)$ is the ith propeller's angular velocity. We assume from the property of the motors (of the propellers) that $\dot{\theta}_i(t) \geq 0$ for all i. Detailed derivation of the above equations of motion can be found in [TOW04].

We can manipulate Equations (16.1), (16.2), (16.3), and (16.4) to become

$$\ddot{\gamma}(t) = C_r\sin\gamma(t) + C_{ur}\left(\dot{\theta}_1^2(t) - \dot{\theta}_3^2(t)\right), \quad (16.5)$$

$$\ddot{\beta}(t) = C_p\sin\beta(t) + C_{up}\left(-\dot{\theta}_2^2(t) + \dot{\theta}_4^2(t)\right), \quad (16.6)$$

$$\ddot{\alpha}(t) = C_{uy}\left(\ddot{\theta}_1(t) - \ddot{\theta}_2(t) + \ddot{\theta}_3(t) - \ddot{\theta}_4(t)\right), \quad (16.7)$$

where C_r, C_p, C_{ur}, C_{up}, and C_{uy} are constants expressed in terms of the parameters M, I_γ, I_β, I_α, I_1, g, e, l_1, l_2, C_L, ρ, and S. Integrating (16.7) with respect to time t yields:

$$\dot{\alpha}(t) - \dot{\alpha}(0) = C_{uy}\left\{\left(\dot{\theta}_1(t) - \dot{\theta}_1(0)\right) - \left(\dot{\theta}_2(t) - \dot{\theta}_2(0)\right)\right.$$
$$\left. + \left(\dot{\theta}_3(t) - \dot{\theta}_3(0)\right) - \left(\dot{\theta}_4(t) - \dot{\theta}_4(0)\right)\right\}, \quad (16.8)$$

where $\dot{\alpha}(0)$, $\dot{\theta}_1(0)$, $\dot{\theta}_2(0)$, $\dot{\theta}_3(0)$, and $\dot{\theta}_4(0)$ are initial angular velocities. Assume that $\dot{\alpha}(0) = 0$. If the starting velocities $\dot{\theta}_1(0) = \dot{\theta}_2(0) = \dot{\theta}_3(0) = \dot{\theta}_4(0)$, (16.8) becomes:

$$\dot{\alpha}(t) = C_{uy}\left(\dot{\theta}_1(t) - \dot{\theta}_2(t) + \dot{\theta}_3(t) - \dot{\theta}_4(t)\right). \quad (16.9)$$

Let ω be an equilibrium point of the propeller's angular velocity. The relation between ω and $\dot{\theta}_i(t)$ is given as

$$\dot{\theta}_i(t) = \omega + \Delta\dot{\theta}_i(t), \qquad (16.10)$$

where $\Delta\dot{\theta}_i(t)$ is the change of the ith propeller's angular velocity about ω. From (16.10), $\Delta\dot{\theta}_i(t) \in [\dot{\theta}_{i\min} - \omega \quad \dot{\theta}_{i\max} - \omega]$, where $\dot{\theta}_{i\max}$ and $\dot{\theta}_{i\min}$ denote the maximum and minimum values of $\dot{\theta}_i(t)$, respectively, with $\dot{\theta}_{i\max} > \dot{\theta}_{i\min} \geq 0$. We note that any ω value satisfying $\max_i \dot{\theta}_{i\min} \leq \omega \leq \min_i \dot{\theta}_{i\max}$, $1 = 1,\ldots,4$, for the four propellers of the helicopter can be selected. Using (16.10), the equations of motion (16.5), (16.6), and (16.9) can be rewritten as follows:

$$
\begin{aligned}
\ddot{\gamma}(t) &= C_r \sin\gamma(t) + C_{ur}\left(2\omega + \Delta\dot{\theta}_1(t) + \Delta\dot{\theta}_3(t)\right) \\
&\quad \times \left(\Delta\dot{\theta}_1(t) - \Delta\dot{\theta}_3(t)\right), \qquad\qquad (16.11) \\
\ddot{\beta}(t) &= C_p \sin\beta(t) + C_{up}\left(2\omega + \Delta\dot{\theta}_2(t) + \Delta\dot{\theta}_4(t)\right) \\
&\quad \times \left(-\Delta\dot{\theta}_2(t) + \Delta\dot{\theta}_4(t)\right), \qquad\qquad (16.12) \\
\dot{\alpha}(t) &= C_{uy}\left(\Delta\dot{\theta}_1(t) + \Delta\dot{\theta}_3(t) - \Delta\dot{\theta}_2(t) - \Delta\dot{\theta}_4(t)\right) \qquad (16.13)
\end{aligned}
$$

Figure 16.1: Picture of the RC helicopter. (Modified from [BKT09]–P. Baranyi, P. Korondi, and K. Tanaka, "Parallel distributed compensation based stabilization of a 3-DOF RC helicopter: A tensor product transformation based approach," *Journal of Advanced Computational Intelligence and Intelligent Informatics*, 13:25–34, 2009.)

Here, the parameters are given as: $C_r = 47.102$ s^{-2}, $C_p = 36.191$ s^{-2}, $C_{ur} = 5.011 \times 10^{-4}$, $C_{up} = 5.126 \times 10^{-4}$, $C_{uy} = 1.155 \times 10^{-3}$, $\dot{\theta}_{i\max} = 2\pi f_0$ rad/s], $\dot{\theta}_{i\min} = 0$ rad/s, $\omega = \pi f_0$ rad/s, where $f_0 = 30$Hz, which is the maximum frequency of the propellers.

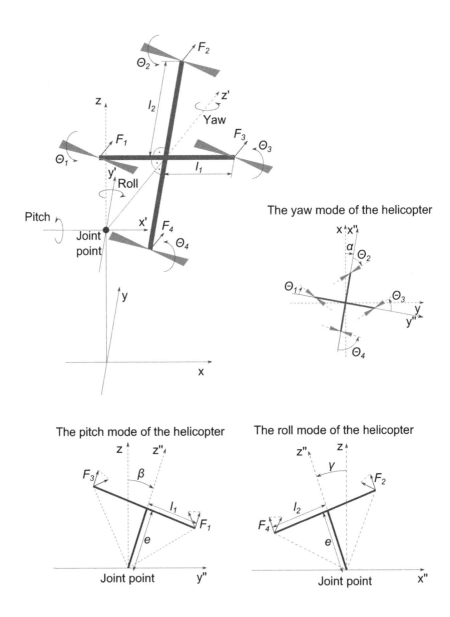

Figure 16.2: Helicopter fixed at a joint point (top left), and as viewed from different perspectives showing the pitch, roll, and yaw modes. (Modified from [BKT09]– P. Baranyi, P. Korondi, and K. Tanaka, "Parallel distributed compensation based stabilization of a 3-DOF RC helicopter: A tensor product transformation based approach," *Journal of Advanced Computational Intelligence and Intelligent Informatics*, 13:25–34, 2009.)

16.1.1 A simplified model

To facilitate control system design, we set to simplify the nonlinear model as obtained above. We consider the assumption that $\Delta\dot\theta_1(t) + \Delta\dot\theta_3(t)$ and $\Delta\dot\theta_2(t) + \Delta\dot\theta_4(t)$ are small. Equations (16.11) and (16.12) can thus be approximated as

$$\ddot\gamma(t) = C_r \sin\gamma(t) + 2C_{ur}\omega\left(\Delta\dot\theta_1(t) - \Delta\dot\theta_3(t)\right), \tag{16.14}$$

$$\ddot\beta(t) = C_p \sin\beta(t) + 2C_{up}\omega\left(-\Delta\dot\theta_2(t) + \Delta\dot\theta_4(t)\right). \tag{16.15}$$

Defining the state vector as

$$\vec{x}(t) = \begin{pmatrix} x_1(t) \\ x_2(t) \\ x_3(t) \\ x_4(t) \\ x_5(t) \end{pmatrix} = \begin{pmatrix} \gamma(t) \\ \dot\gamma(t) \\ \beta(t) \\ \dot\beta(t) \\ \alpha(t) \end{pmatrix}$$

and the input vector as

$$\vec{u}(t) = \begin{pmatrix} u_1(t) \\ u_2(t) \\ u_3(t) \\ u_4(t) \end{pmatrix} = \begin{pmatrix} \Delta\dot\theta_1(t) - \Delta\dot\theta_3(t) \\ \Delta\dot\theta_1(t) + \Delta\dot\theta_3(t) \\ -\Delta\dot\theta_2(t) + \Delta\dot\theta_4(t) \\ \Delta\dot\theta_2(t) + \Delta\dot\theta_4(t) \end{pmatrix} \approx \begin{pmatrix} \Delta\dot\theta_1(t) - \Delta\dot\theta_3(t) \\ 0 \\ -\Delta\dot\theta_2(t) + \Delta\dot\theta_4(t) \\ 0 \end{pmatrix} \tag{16.16}$$

then yields the following simplified quasi-linear parameter-varying (qLPV) model for the RC helicopter:

$$\dot{\mathbf{x}}(t) = \mathbf{A}(\mathbf{p}(t))\mathbf{x}(t) + \mathbf{B}\mathbf{u}(t), \tag{16.17}$$

where $\mathbf{p}(t) = \begin{pmatrix} x_1(t) & x_3(t) \end{pmatrix}^T = \begin{pmatrix} \gamma(t) & \beta(t) \end{pmatrix}^T$.

16.1.2 Modeling of uncertainty

The model (16.17) is a simplification of the original Equations (16.11) to (16.13). Hence, the controller design based on (16.17) may not always stabilize the original nonlinear dynamics. Some sort of robust control design must therefore be employed. Here, we derive an uncertainty description to cover the modeling discrepancies between the simplified and original dynamics.

It can be observed from (16.16) that the modeling error is

$$\begin{bmatrix} 0 & 0 & 0 & 0 \\ C_{ur}u_2(t) & 0 & 0 & 0 \\ 0 & 0 & 0 & 0 \\ 0 & 0 & C_{up}u_4(t) & 0 \\ 0 & 0 & 0 & 0 \end{bmatrix} \mathbf{u}(t). \tag{16.18}$$

With (16.18), one can obtain a qLPV model for the helicopter incorporating un-certainty elements as follows:

$$\dot{\mathbf{x}}(t) = (\mathbf{A}(\mathbf{p}(t)) + \mathbf{D}_a(\mathbf{p}(t))\Delta_a(t)\mathbf{E}_a(\mathbf{p}(t)))) \mathbf{x}(t) \tag{16.19}$$

$$+ (\mathbf{B}(\mathbf{p}(t)) + \mathbf{D}_b(\mathbf{p}(t))\Delta_b(t)\mathbf{E}_b(\mathbf{p}(t)))) \mathbf{u}(t).$$

Generally, $\mathbf{D}_a$, $\mathbf{E}_a$, $\mathbf{D}_b$, and $\mathbf{E}_b$ are known matrices, and $\Delta_a(t)$ and $\Delta_b(t)$ are unknown uncertain blocks. In the present case, we have

$$\Delta_a(t) = 0, \ \mathbf{D}_a = 0, \ \mathbf{E}_a = 0,$$

$$\Delta_b(t) = \begin{bmatrix} u_2(t) & 0 & 0 & 0 \\ 0 & 0 & 0 & 0 \\ 0 & 0 & u_4(t) & 0 \\ 0 & 0 & 0 & 0 \end{bmatrix}, \tag{16.20}$$

$$\mathbf{D}_b = \begin{bmatrix} 0 & 0 & 0 & 0 \\ C_{ur} & 0 & 0 & 0 \\ 0 & 0 & 0 & 0 \\ 0 & 0 & C_{up} & 0 \\ 0 & 0 & 0 & 0 \end{bmatrix}, \ \mathbf{E}_b = \mathbf{I}_4.$$

Particularly, the upper bounds $\frac{1}{\rho_a}$ and $\frac{1}{\rho_b}$ of the uncertain blocks $\|\Delta_a(t)\|$ and $\|\Delta_b(t)\|$ are known in this case, that is, $\|\Delta_a(t)\| \leq \frac{1}{\rho_a} = 0$ and $\|\Delta_b(t)\| \leq \frac{1}{\rho_b} = \sqrt{u_{2\max}^2 + u_{4\max}^2}$, where $u_{2\max}$ and $u_{4\max}$ denote the maximum values of $|u_2(t)|$ and $|u_4(t)|$, respectively. These values correspond to the saturations of the actuators (i.e., the angular velocity of the propellers).

Note that in order to accommodate qLPV models with uncertainty description such as (16.19), the TP transformation will need to be augmented with the following terms:

$$\mathbf{D}_{a/b}(\mathbf{p}(t)) = \mathcal{D}_{a/b} \underset{n=1}{\overset{N}{\boxtimes}} \mathbf{w}_n(p_n(t))\mathbf{x}(t), \tag{16.21}$$

$$\mathbf{E}_{a/b}(\mathbf{p}(t)) = \mathcal{E}_{a/b} \underset{n=1}{\overset{N}{\boxtimes}} \mathbf{w}_n(p_n(t))\mathbf{x}(t). \tag{16.22}$$

16.2   The TP model

We set the transformation space Ω as $x_1(t) \in [-\pi/2, \pi/2]$ and $x_3(t) \in [-\pi/2, \pi/2]$, that is, $\Omega : [-\pi/2, \pi/2] \times [-\pi/2, \pi/2]$, and adopt the equidistant discretization hyper-grid of $M_1 \times M_2 : 301 \times 301$. In this regard, as the transformation space is symmetric about the stabilization point for both x_1 and x_2, we set an odd grid size along both directions to ensure that the stabilization point will lie exactly on one of the grid

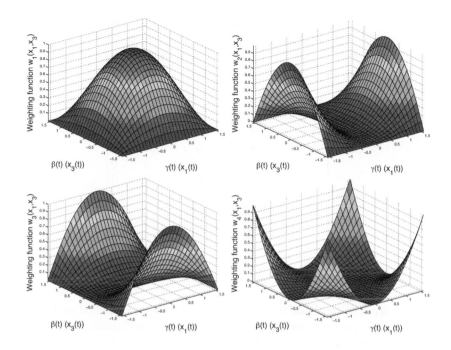

Figure 16.3: Weighting functions $w_r(x_1, x_2)$ of roll angle $x_1(t)$ and pitch angle $x_3(t)$.

points. We also generate the CNO type weighting functions to yield a tight convex hull. The TP model transformation results in

$$\dot{\mathbf{x}}(t) = \left(\sum_{r=1}^{4} w_r(\mathbf{p}(t))\mathbf{A}_r \right) \mathbf{x}(t) + (\mathbf{B} + \mathbf{D}_b \delta_b \mathbf{E}_b) \mathbf{u}(t). \tag{16.23}$$

The weighting functions $w_r(\mathbf{p}(t))$, $r = 1, \ldots, 4$ are depicted in Figure 16.3. We can observe that one linear time-invariant (LTI) system dominates in the equilibrium point and the others on the border of Ω.

The resulting LTI systems are:

$$\mathbf{A}_1 = \begin{pmatrix} 0 & 1.0000 & 0 & 0 & 0 \\ 47.1020 & 0 & 0 & 0 & 0 \\ 0 & 0 & 0 & 1.0000 & 0 \\ 0 & 0 & 36.1910 & 0 & 0 \\ 0 & 0 & 0 & 0 & 0 \end{pmatrix},$$

$$\mathbf{A}_2 = \begin{pmatrix} 0 & 1.0000 & 0 & 0 & 0 \\ 29.9861 & 0 & 0 & 0 & 0 \\ 0 & 0 & 0 & 1.0000 & 0 \\ 0 & 0 & 36.1910 & 0 & 0 \\ 0 & 0 & 0 & 0 & 0 \end{pmatrix},$$

$$\mathbf{A}_3 = \begin{pmatrix} 0 & 1.0000 & 0 & 0 & 0 \\ 47.1020 & 0 & 0 & 0 & 0 \\ 0 & 0 & 0 & 1.0000 & 0 \\ 0 & 0 & 23.0399 & 0 & 0 \\ 0 & 0 & 0 & 0 & 0 \end{pmatrix},$$

$$\mathbf{A}_4 = \begin{pmatrix} 0 & 1.0000 & 0 & 0 & 0 \\ 29.9861 & 0 & 0 & 0 & 0 \\ 0 & 0 & 0 & 1.0000 & 0 \\ 0 & 0 & 23.0399 & 0 & 0 \\ 0 & 0 & 0 & 0 & 0 \end{pmatrix}.$$

Hence, it can be concluded from the above results that the simplified model of the 3-DoF RC helicopter incorporated with uncertainty elements can exactly be given by a convex combination of four LTI systems, which in turn define the tight convex hull of the given qLPV model. This fact can actually be derived analytically from the model Equation (16.19), but the TP model transformation was able to come up with such a result in a few minutes of computation independently of the actual analytical form of the given model.

16.3   Control system design

Let us first outline the design specifications for our control system:

- We desire a robust controller, as the simplified model used for design constitutes only an approximation of the real situation and omits some of the real dynamics, for example, ground effect and propeller deflection.

- We desire decay rate control to impose an adequately speedy system response.

- We like to impose constraints on the control value according to the operation range of the propeller engines.

- We assume the initial state vector is unknown.

- The controller must take into account the uncertain components of the qLPV model above.

In order to meet the robustness specifications above, we note that the simplified uncertain model of the RC helicopter (16.23) has $\Delta_a = 0$, $\mathbf{D}_a = 0$, $\mathbf{E}_a = 0$, and that $\mathbf{B}$, Δ_b, $\mathbf{D}_b$ and $\mathbf{E}_b$ do not depend on $\mathbf{p}(t) \in \Omega$ at all. We recall the following linear matrix inequality (LMI) from [TOW04] that guarantees robust stability and decay rate condition:

$$\underset{\mathbf{X}, \mathbf{M}_r, d_b}{\text{maximize}} \quad \alpha \quad \text{subject to} \tag{16.24}$$

$$\text{subject to} \quad \mathbf{X} > \mathbf{0}, \quad d_b > 0, \quad \hat{\mathbf{S}}_{rr} < \mathbf{0},$$

where

$$\hat{\mathbf{S}}_{rr} = \begin{bmatrix} \left(\mathcal{L}(\mathbf{A}_r, \mathbf{B}, \mathbf{X}, \mathbf{M}_r) + 2\zeta\mathbf{X} + d_b\mathbf{D}_b\mathbf{D}_b^T \right) & * \\ -\mathbf{E}_b\mathbf{M}_r & -d_b\rho_b^2\mathbf{I} \end{bmatrix}$$

and

$$\mathcal{L}(\mathbf{A}_r, \mathbf{B}_r, \mathbf{X}, \mathbf{M}_s) = \mathbf{X}\mathbf{A}_r^T + \mathbf{A}_r\mathbf{X} - \mathbf{B}_r\mathbf{M}_s - \mathbf{M}_s^T\mathbf{B}_r^T.$$

for $r < s \leq R$, except for the pairs (r, s) such that $\forall \mathbf{p}(t) : w_r(\mathbf{p}(t))w_s(\mathbf{p}(t)) = 0$. The symbol $*$ denotes the transposed elements (matrices) for symmetric positions.

Moreover, in order to impose upper limits on the control values, we further adopt the LMI condition from Theorem 11.7. Assume that the initial $\|x(0)\| \leq \phi$, where $x(0)$ is unknown but its upper bound ϕ is known. The constraints $|u_j(t)| \leq \mu_j$, $j = 1, 2, \ldots, m$ are then guaranteed at all times $t \geq 0$ if there exist a common positive definite matrix $\mathbf{X}$ and matrices $\mathbf{M}_r$ such that

$$\phi^2 I \leq \mathbf{X}, \quad \begin{pmatrix} \mathbf{X} & \mathbf{M}_r^T\mathbf{E}_j^T \\ \mathbf{E}_j\mathbf{M}_r & \mu_j^2\mathbf{I} \end{pmatrix} \geq \mathbf{0},$$

where $\mathbf{E}_j$ is the vector whose jth element is one and the other elements are zero, that is,

$$\begin{array}{cccccccc} & 1st & & (j\text{-}1)th & jth & (j\text{+}1)th & & mth \\ \mathbf{E}_j = [& 0 & \cdots & 0 & 1 & 0 & \cdots & 0 &]. \end{array}$$

Note that matrix $\mathbf{E}_j$ allows each input variable to be described as $u_j(t) = \mathbf{E}_j\mathbf{u}(t)$, where $\mathbf{u}(t) = [u_1(t) \ u_2(t) \ \cdots \ u_m(t)]^T$.

The controller with the desired performance is obtained through solving LMI conditions (16.24) and (16.25) above. The solution yields feedback gains $\mathbf{F}_r$, $r = 1, \ldots, R$ as

$$\mathbf{F}_r = \mathbf{M}_r\mathbf{X}^{-1}. \tag{16.25}$$

The control value is computed by (11.19) as

$$\mathbf{u}(t) = -\left(\sum_{r=1}^{4} w_r(\mathbf{p}(t))\mathbf{F}_r \right) \mathbf{x}(t). \tag{16.26}$$

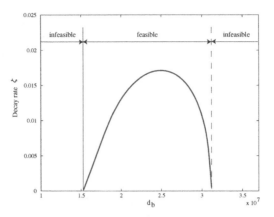

Figure 16.4: Decay rate ζ for d_b.

In short, we can obtain the feedback gains satisfying the above multiobjectives of robust stability with good decay rate and input constraints through TP model-based design and properly selected LMI conditions.

16.4   Control performance

The parameters on the input constraints utilized in our design are

$$\mathbf{E}_1 = [1 \ 0 \ 0 \ 0], \quad \mathbf{E}_2 = [0 \ 1 \ 0 \ 0],$$
$$\mathbf{E}_3 = [0 \ 0 \ 1 \ 0], \quad \mathbf{E}_4 = [0 \ 0 \ 0 \ 1],$$

and

$$\mu_1 = \mu_3 = 0.9 \times 2\pi f_0, \quad \mu_2 = \mu_4 = 0.1 \times 2\pi f_0.$$

Figure 16.4 depicts the change of decay rate ζ as a function of d_b. The LMI conditions (16.24) and (16.25) are feasible in the range from d_b ranging from 1.524×10^7 to 3.119×10^7. For the present design, we select $d_b = 2.495 \times 10^7$ to yield the maximum decay rate.

The resulting LTI feedback systems obtained accordingly are

$$\mathbf{F}_1 = \begin{pmatrix} 569.3183 & 82.9636 & -32.2101 & -5.3726 & 0.3969 \\ 0.3337 & 0.0486 & -0.0188 & -0.0031 & 15.8392 \\ 344.5022 & 50.5279 & 694.2790 & 114.5536 & 16.7862 \\ -0.3337 & -0.0486 & 0.0188 & 0.0031 & -15.8392 \end{pmatrix},$$

$$\mathbf{F}_2 = \begin{pmatrix} 698.6409 & 103.1000 & 97.1139 & 18.0629 & 6.0978 \\ 3.4361 & 0.5149 & 1.2268 & 0.2340 & 15.9510 \\ 322.4559 & 52.2701 & 588.9798 & 113.3114 & 15.5008 \\ -3.4361 & -0.5149 & -1.2268 & -0.2340 & -15.9510 \end{pmatrix},$$

$$\mathbf{F}_3 = \begin{pmatrix} 644.6347 & 102.6238 & 125.1829 & 17.4642 & 12.2166 \\ 0.8747 & 0.1391 & 0.3364 & 0.0613 & 12.7815 \\ 315.4957 & 49.4383 & 672.3149 & 111.2250 & 14.4206 \\ -0.8747 & -0.1391 & -0.3364 & -0.0613 & -12.7815 \end{pmatrix},$$

$$\mathbf{F}_4 = \begin{pmatrix} 526.2796 & 82.5688 & -1.5450 & -5.8320 & 6.8792 \\ 10.3008 & 1.6208 & 0.6431 & 0.0280 & 13.0706 \\ 256.3282 & 44.8857 & 578.9468 & 111.5533 & 12.6039 \\ -10.3008 & -1.6208 & -0.6431 & -0.0280 & -13.0706 \end{pmatrix}.$$

These $\mathbf{F}_r$ quantities allow the control function to be expressed as (11.19). Figure 16.5 shows the closed-loop controlled results of the 3-DoF RC helicopter. One can observe that the decay rate condition, the robust stability condition, and the input constraint condition are all satisfied. In this case, we set the initial state values as

$$\mathbf{x}(0) = [0.279 \quad 0 \quad -0.279 \quad 0 \quad -0.349]^T.$$

For comparison, we also designed a controller guaranteeing only the decay rate condition but without the robust stability condition. The resulting control performance is presented in Figure 16.6. One can observe that the robust controller shows better performance in terms of settling time.

Remark 16.1. Actual operation of the helicopter is influenced by ground effect and the loss of lift power due to deflection of the propellers when rotating at high frequency. As these two factors were not considered in our modeling step, the control performance of the helicopter in a real experiment may actually be different from what we have here. In this regard, it is worth mentioning that the helicopter controllers derived here are equivalent to those obtained in [TOW04]. Thus, one should expect the derived controller here would yield the same experimental results as those observed in the successful laboratory experiment detailed in [TOW04].

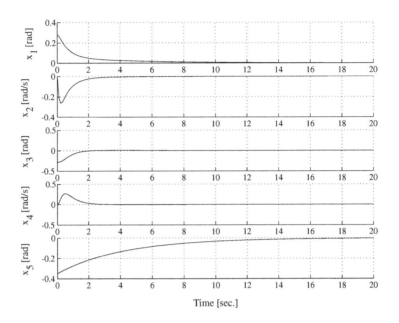

Figure 16.5: Control result 1 (robust stability & decay rate).

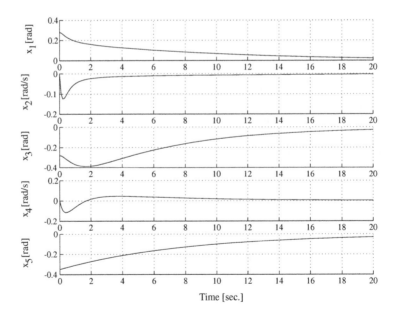

Figure 16.6: Control result 2 (decay rate controller).

Chapter 17

Heavy Vehicle Rollover Prevention Problem

This chapter is based largely on the works of Nagy, Baranyi, and Gáspár [NBG08] on the rollover prevention of heavy vehicles and their ability to resist overturning moments generated during cornering. A combined yaw–roll model including the roll dynamics of unsprung masses is first presented in this chapter. The model is non-linear with respect to the velocity of the vehicle, which is handled as a quasi-linear parameter-varying (qLPV) scheduling parameter here. The qLPV model of the vehicle is converted into a proper polytopic form by tensor product (TP) model transformation. $\mathcal{H}_\infty$ gain scheduling is then applied for stabilization control design. Numerical simulation is also included to demonstrate the effectiveness of the designed controller.

17.1   Problem introduction

Roll stability is determined by the height of the center of mass, the track width, and the kinematic properties of the suspension of a vehicle. Heavy vehicles suffer from a relatively high mass center and narrow track width, and are hence more susceptible to rollover. This is especially so when the vehicle is changing lanes or trying to avoid obstacles. The vehicle body rolls out of the corner and the center of mass shifts outboard of the centerline, creating a destabilizing moment.

Many papers on different approaches to the active control of vehicle dynamics to reduce the risk of rollover can be found in the literature. These works can largely be

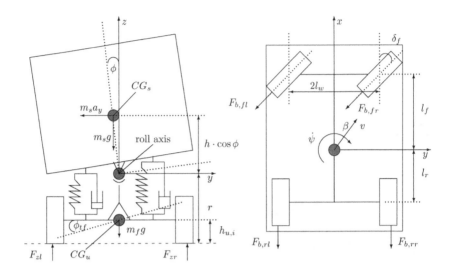

Figure 17.1: Rollover vehicle model. (Replicated with permission [NBG08]–S. Nagy, P. Baranyi, and P. Gáspár, "Rollover prevention of a heavy vehicle via TP model based H-infinity control design approach," *Acta Technica Jauriensis, Series Intelligentia Computatorica*, 1(3):531–546, December 2008.)

classified into three major schemes of active intervention into the vehicle dynamics, namely, active antiroll bars, active steering, and active braking. Their controller designs are usually based on linear time-invariant (LTI) models and linear approaches, with the forward velocity usually handled as a constant parameter in the yaw–roll model.

In this chapter, we adopt a combined yaw–roll model that includes the roll dynamics of unsprung masses [GSB04]. Different from previous works, our model is nonlinear with respect to the vehicle velocity, which is handled as a qLPV scheduling parameter here. The idea is thus to adjust the controller continuously according to the measurement of the vehicle velocity in real time. This is in line with the works of [GSB04], [GSB02], [PSG99] showing that the velocity is actually an important parameter relevant to roll stability.

17.2 A qLPV model for heavy vehicles

Figure 17.1 models the combined yaw–roll dynamics of the vehicle as a three-body system, in which m_s is the sprung mass, $m_{u,f}$ is the unsprung mass at the front, and

$m_{u,r}$ the unsprung mass at the rear. Note that $m_{u,f}$ includes also the wheel and axles at the front, and $m_{u,r}$ the same at the rear.

The conditions of the yaw–roll model used in control design are now discussed. It is assumed that the roll axis runs along the longitudinal direction of the vehicle lying parallel to the road at a height r above the road. The exact location of the roll axis depends on the kinematic properties of the front and rear suspension. The axles of the vehicle are considered to be a single rigid body with flexible tires that can rotate about the center of the roll. The tire characteristics in the model are assumed to be linear. The effect caused by pitching dynamics in the longitudinal plane can be ignored in the present modeling of the vehicle behavior. The effects of aerodynamic inputs (wind disturbance) and road disturbances are also ignored. The roll motion of the sprung mass is damped by the suspension and stabilizers of effective roll damping coefficients $b_{s,i}$ and roll stiffness $k_{s,i}$.

The lateral dynamics, the yaw moment, and the roll moment of the sprung and the unsprung masses are taken into consideration in the modeling process. The symbols of the yaw–roll model are listed in Table 17.1. The motion differential equations are shown in the following:

$$mv(\dot{\beta} + \dot{\psi}) - m_s h \ddot{\phi} = Y_\beta \beta + Y_\psi \dot{\psi} + Y_{\delta_f} \delta_f \tag{17.1}$$

$$-I_{xz}\ddot{\phi} + I_{zz}\ddot{\psi} = N_\beta \beta + N_\psi \dot{\psi} + N_{\delta_f} \delta_f + \frac{l_w}{2}\Delta F_b \tag{17.2}$$

$$\left(I_{xx} + m_s h^2\right)\ddot{\phi} - I_{xz}\ddot{\psi} = m_s g h \phi + m_s v h (\dot{\beta} + \dot{\psi}) - k_f(\phi - \phi_{t,f}) - b_f(\dot{\phi} - \dot{\phi}_{t,f})$$
$$- k_r(\phi - \phi_{t,r}) - b_r(\dot{\phi} - \dot{\phi}_{t,r}) \tag{17.3}$$

$$-r\left(Y_{\beta,f}\beta + Y_{\psi,f}\dot{\psi} + Y_{\delta_f}\delta_f\right) = m_{u,f}v(r - h_{u,f})(\dot{\beta} + \dot{\phi}) + m_{u,f}gh_{u,f}\phi_{t,f} - k_{t,f}\phi_{t,f}$$
$$+ k_f(\phi - \phi_{t,f}) + b_f(\dot{\phi} - \dot{\phi}_{t,f}) \tag{17.4}$$

$$-r\left(Y_{\beta,r}\beta + Y_{\psi,r}\dot{\psi}\right) = m_{u,r}v(r - h_{u,r})(\dot{\beta} + \dot{\psi}) - m_{u,r}gh_{u,r}\phi_{t,r} - k_{t,r}\phi_{t,r}$$
$$+ k_r(\phi - \phi_{t,r}) + b_r(\dot{\phi} - \dot{\phi}_{t,r}) \tag{17.5}$$

Here, the tire coefficients are given by: $Y_\beta = -(C_f + C_r)\mu$, $N_\beta = (C_r l_r - C_f l_f)\mu$, $Y_\psi = (C_r l_r - C_f l_f)\frac{\mu}{v}$, $N_\psi = -(C_f l_f^2 + C_r l_r^2)\frac{\mu}{v}$, $Y_{\delta_f} = C_f\mu$, $N_{\delta_f} = C_f l_f \mu$. These equations can be expressed in a state space representation. Let the state vector be the following:

$$\mathbf{x}(t) = \begin{bmatrix} \beta & \dot{\psi} & \phi & \dot{\phi} & \phi_{t,f} & \phi_{t,r} \end{bmatrix}^T. \tag{17.6}$$

The system states are the side-slip angle of the sprung mass β, the yaw rate $\dot{\psi}$, the roll angle ϕ, the roll rate $\dot{\phi}$, and the roll angles $\phi_{t,f}$ and $\phi_{t,r}$ of the unsprung mass at the front and rear axles, respectively. Then the state equation can be given in the following form:

$$\mathbf{E}(\mathbf{p}(t))\dot{\mathbf{x}}(t) = \mathbf{A}_0(\mathbf{p}(t))\mathbf{x}(t) + \mathbf{B}_{1,0}\delta_f + \mathbf{B}_{2,0}u)(t), \tag{17.7}$$

where the matrices are defined by the following equations:

$$
\mathbf{E}(v) = \begin{bmatrix}
mv & 0 & 0 & -m_sh & 0 & 0 \\
0 & I_{zz} & 0 & -I_{xz} & 0 & 0 \\
-m_svh & -I_{xz} & 0 & I_{xx}+m_sh^2 & -b_f & -b_r \\
m_{u,f}v(r-h_{u,f}) & 0 & 0 & 0 & -b_f & 0 \\
m_{u,r}v(r-h_{u,r}) & 0 & 0 & 0 & 0 & -b_r \\
0 & 0 & 1 & 0 & 0 & 0
\end{bmatrix},
\tag{17.8}
$$

$$
\mathbf{B}_{1,0} = \begin{bmatrix} Y_{\delta_f} \\ N_{\delta_f} \\ 0 \\ rY_{\delta_f} \\ 0 \\ 0 \end{bmatrix}, \qquad
\mathbf{B}_{2,0} = \begin{bmatrix} 0 \\ l_w/2 \\ 0 \\ 0 \\ 0 \\ 0 \end{bmatrix}
\tag{17.9}
$$

$$
\mathbf{A}_0(v) = \begin{bmatrix}
Y_\beta & Y_{\dot\psi}-mv & 0 & 0 & 0 & 0 \\
N_\beta & N_{\dot\psi} & 0 & 0 & 0 & 0 \\
0 & m_shv & A_{33} & -b_f-b_r & k_f & k_r \\
-rY_{\beta,f} & A_{42} & -k_f & -b_f & A_{45} & 0 \\
-rY_{\beta,r} & A_{52} & -k_r & -b_r & 0 & A_{56} \\
0 & 0 & 0 & 1 & 0 & 0
\end{bmatrix}
\tag{17.10}
$$

where $A_{33} = (m_sgh-k_f-k_r)$, $A_{42} = rY_{\dot\psi,f}-m_{u,f}v(r-h_{u,f})$, $A_{45} = (k_f+k_{t,f}-m_{u,f}ghu,f)$, $A_{52} = -rY_{\dot\psi,r} - m_{u,r}v(r-h_{u,r}))$, and $A_{56} = (k_r + k_{t,r} - m_{u,r}ghu,r)$.

The parameter of the system is the forward velocity

$$\mathbf{p} = v.$$

Equation (17.7) can be further rewritten as

$$\dot{\mathbf{x}}(t) = \mathbf{A}(\mathbf{p}(t))\mathbf{x}(t) + \mathbf{B}_1(\mathbf{p}(t))\delta_f + \mathbf{B}_2(\mathbf{p}(t))u(t), \tag{17.11}$$

where

$$\mathbf{A}(\mathbf{p}(t)) = \mathbf{E}^{-1}(\mathbf{p}(t))\mathbf{A}_0(\mathbf{p}(t)) \tag{17.12}$$

$$\mathbf{B}_1(\mathbf{p}(t)) = \mathbf{E}^{-1}(\mathbf{p}(t))\mathbf{B}_{1,0} \tag{17.13}$$

$$\mathbf{B}_2(\mathbf{p}(t)) = \mathbf{E}^{-1}(\mathbf{p}(t))\mathbf{B}_{2,0} \tag{17.14}$$

Here, δ_f is the front wheel steering angle. The control input is the difference of brake forces between the left and the right hand side of the vehicle,

$$u = \Delta F_b.$$

Table 17.1: Symbols of the Yaw–Roll Model

Symbols	Description
h	height of CG of sprung mass from roll axis
$h_{u,i}$	height of CG of unsprung mass from ground
r	height of roll axis from ground
a_y	lateral acceleration
β	side-slip angle at center of mass
ψ	heading angle
$\dot{\psi}$	yaw rate
ϕ	sprung mass roll angle
$\phi_{t,i}$	unsprung mass roll angle
δ_f	steering angle
u_i	control torque
C_i	tire cornering stiffness
F_{zi}	total axle load
R_i	normalized load transfer
k_i	suspension roll stiffness
b_i	suspension roll damping
$k_{t,i}$	tire roll stiffness
I_{xx}	roll moment of inertia of sprung mass
I_{xz}	yaw–roll product of inertia of sprung mass
I_{zz}	yaw moment of inertia of sprung mass
l_i	length of the axle from the CG
l_w	vehicle width
μ	road adhesion coefficient

The control input provided by the brake system generates a yaw moment which affects the lateral tire forces directly. In our case, it is further assumed that the brake force difference ΔF_b provided by the controller is applied to the rear axle, meaning that only one wheel is decelerated at the rear axle. This deceleration can be caused by an appropriate yaw moment. In our case, the difference between the brake forces can be given by $\Delta F_b = F_{b,rl} - F_{b,rr}$. This assumption does not restrict the implementation of the controller because it is possible that the control action be distributed on the front and the rear wheels at one of the two sides. The reason for distributing the control force to the front and the rear wheels is to minimize the wear of the tires. This requires a logic which calculates the brake forces for the wheels.

In Equation (17.11), the matrix $\mathbf{A}(\mathbf{p}(t))$ depends on the forward velocity of the vehicle nonlinearly. As mentioned, this contrasts with the linear yaw–roll models adopted in previous works, where the velocity is considered a constant. Moreover, with the throttle kept constant during lateral maneuvers, the forward velocity will hence depend only on the brake forces. The differential equation for forward velocity is

$$m\dot{v} = -F_{b,rl} - F_{b,rr}.$$

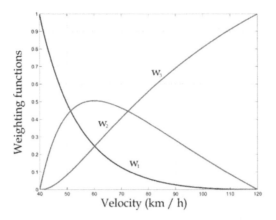

Figure 17.2: CNO weighting functions of the TP model.

17.3   The TP model

The TP model of the qLPV model (17.11) is obtained by executing the TP model transformation procedures over the v-domain of $\Omega = [40\text{km/h}, 120\text{km/h}]$ with $M = 137$. The process shows that the qLPV model of the heavy vehicle can be exactly given by the convex combination of three LTI vertex systems:

$$\mathbf{S}(\mathbf{p}(t)) = \sum_{i=1}^{3} w_i(\mathbf{p}(t))\mathbf{S}_i, \qquad (17.15)$$

where we adopted close-to-normality (CNO) type weighting functions to define a tight convex so as to relax the feasibility of the possible linear matrix inequality (LMI) conditions. The resulting weightings are depicted in Figure 17.2.

17.4   Control system design and performance

The aim of the rollover prevention is to provide the vehicle with the ability to re-sist overturning moments generated during cornering. We utilize the convex model obtained above for the stabilization control design of the heavy vehicle. We seek a qLPV controller of the form

$$\begin{aligned} \dot{\mathbf{x}}_K &= \mathbf{A}_K(\mathbf{p}(t))\mathbf{x}_K + \mathbf{B}_K(\mathbf{p}(t))y \\ u &= \mathbf{C}_K(\mathbf{p}(t))\mathbf{x}_K + \mathbf{D}_K(\mathbf{p}(t))y, \end{aligned}$$

where

$$\begin{pmatrix} \mathbf{A}_K(\mathbf{p}(t)) & \mathbf{B}_K(\mathbf{p}(t)) \\ \mathbf{C}_K(\mathbf{p}(t)) & \mathbf{D}_K(\mathbf{p}(t)) \end{pmatrix} = \mathbf{K}(\mathbf{p}(t)) = \sum_{i=1}^{3} w_i(\mathbf{p}(t))\mathbf{K}_i$$

with the same $w_i(\mathbf{p}(t))$ weighting functions as in the model representation (17.15). $\mathbf{x}_K$ is the internal state of the controller. The measured output of the model is the yaw rate, $y = \dot{\psi}$, and $u = \Delta F_b$ is the control signal. To design a suitable $\mathbf{K}(\mathbf{p})$ for the given polytopic model, the self-scheduled $\mathcal{H}_\infty$ controller design method [AGB94], [AG95] was used.

The closed-loop interconnection structure, which includes the feedback structure of the model $\mathbf{P}$ and controller $\mathbf{K}$, is shown in Figure 17.3. In the figure, $\mathbf{d}$, $\mathbf{u}$, $\mathbf{y}$, and $\mathbf{z}$ are the disturbance, the control input, the measured output, and the performance output respectively.

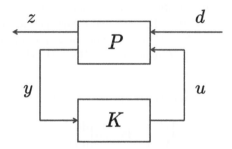

Figure 17.3: The general *P-K* structure for control design. (Replicated with permission from [NBG08]–S. Nagy, P. Baranyi, and P. Gáspár, "Rollover prevention of a heavy vehicle via TP model based H-infinity control design approach," *Acta Technica Jauriensis, Series Intelligentia Computatorica*, 1(3):531–546, December 2008.)

A standard feedback configuration with additional weightings is illustrated in Figure 17.4. In the figure, $\mathbf{u}$ is the control input, $\mathbf{y}$ is the measured output, $\mathbf{z}_p$ is the performance output, $\mathbf{z}_u$ and $\mathbf{z}_y$ are performances at the input and the output, $\mathbf{w}$ is the disturbance, and $\mathbf{n}$ is the measurement noise. The purpose of the weighting function $\mathbf{W}_p$ is to define the performance specifications. They can be considered as penalty functions, that is, weightings should be heavy in a frequency range where small signals are desired, and light where large performance outputs can be tolerated. $\mathbf{W}_u$ and $\mathbf{W}_y$ may be used to reflect some restrictions on the actuator and on the output signals. The purpose of the weighting functions $\mathbf{W}_w$ and $\mathbf{W}_n$ is to reflect the disturbance and sensor noises. The disturbance and the performance in the general $\mathbf{P}$-$\mathbf{K}$ structure are $\mathbf{d} = \begin{bmatrix} \mathbf{w} & \mathbf{n} \end{bmatrix}^T$ and $\mathbf{z} = \begin{bmatrix} \mathbf{z}_u & \mathbf{z}_y & \mathbf{z}_p \end{bmatrix}^T$.

The augmented plant that includes the parameter-dependent vehicle dynamics

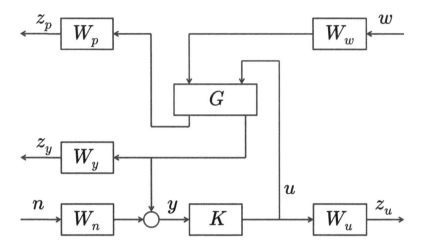

Figure 17.4: The standard feedback configuration with weights. (Replicated with permission [NBG08]–S. Nagy, P. Baranyi, and P. Gáspár, "Rollover prevention of a heavy vehicle via TP model based H-infinity control design approach," *Acta Technica Jauriensis, Series Intelligentia Computatorica*, 1(3):531–546, December 2008.)

and the weighting functions are thus defined in the following form:

$$\begin{bmatrix} \mathbf{z}(t) \\ \mathbf{y}(t) \end{bmatrix} = \mathbf{P}(\mathbf{p}(t)) \begin{bmatrix} \mathbf{d}(t) \\ \mathbf{u}(t) \end{bmatrix}. \tag{17.16}$$

The closed-loop system $\mathbf{M}(\mathbf{p}(t))$ is given by a lower linear fractional transformation structure:

$$\mathbf{M}(\mathbf{p}(t)) = \mathcal{F}_\ell(\mathbf{P}(\mathbf{p}(t)), \mathbf{K}(\mathbf{p}(t))), \tag{17.17}$$

where $\mathbf{K}(\mathbf{p}(t))$ also depends on the scheduling variable $\mathbf{p}(t)$. The goal of the control design is to minimize the induced $\mathcal{L}_2$ norm of a qLPV system $\mathbf{M}(\mathbf{p}(t))$, with zero initial conditions. This is given by

$$\|\mathbf{M}(\mathbf{p}(t))\|_\infty = \sup_{\mathbf{p}(t)\in\Omega} \sup_{\|\mathbf{w}\|_2\neq0, \mathbf{w}\in\mathcal{L}_2} \frac{\|\mathbf{z}(t)\|_2}{\|\mathbf{w}(t)\|_2}. \tag{17.18}$$

We now present a simulation case study. At the beginning, let the velocity of the vehicle be $v = 120$km/h, and all other state variables be zero. Then a sharp maneuver is simulated via the steering angle δ_f corresponding to the situation when the vehicle has to perform an obstacle avoidance maneuver, as depicted in Figure 17.5. The control goal is to stabilize the vehicle by braking the rear wheels. Figure 17.6 shows the closed-loop results. It can be observed that the likelihood of rollover can be reduced efficiently with the active braking control generated.

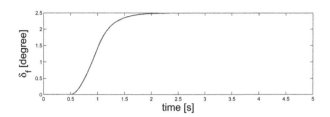

Figure 17.5: Disturbance signal: Steering angle for simulation case study.

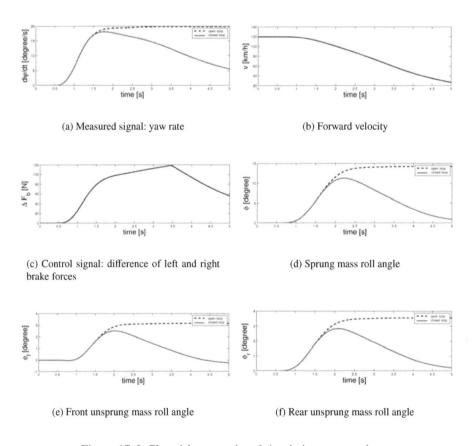

(a) Measured signal: yaw rate

(b) Forward velocity

(c) Control signal: difference of left and right brake forces

(d) Sprung mass roll angle

(e) Front unsprung mass roll angle

(f) Rear unsprung mass roll angle

Figure 17.6: Closed-loop results of simulation case study.

References

[AG95]   P. Apkarian and P. Gahinet.   A convex characterization of gain-scheduled H_∞ controllers. *IEEE Transactions on Automatic Control*, 40(5):853–864, 1995.

[AGB94]   P. Apkarian, P. Gahinet, and G. Becker. Self-scheduled H_∞ linear parameter-varying systems. In *Proceedings of the 1994 American Control Conference*, volume 1, pages 856–860, Baltimore, Maryland, USA, 1994.

[AHJM12]   B.M. Al-Hadithi, A. Jiménez, and F. Matía. New methods for the estimation of Takagi–Sugeno model based extended Kalman filter and its applications to optimal control for nonlinear systems. *Optimal Control Applications and Methods*, 33(5):552–575, September 2012.

[Arn57]   V.I. Arnold. On functions of three variables. In *Doklady Akademii Nauk SSSR*, volume 114, pages 679–681, 1957.

[Bar83]   B.R. Barmish. Stabilization of uncertain systems via linear control. *IEEE Transactions on Automatic Control*, 28(8):848–850, 1983.

[Bar04]   P. Baranyi. TP model transformation as a way to LMI-based controller design. *IEEE Transactions on Industrial Electronics*, 51(2):387–400, 2004.

[Bar06a]   P. Baranyi. Output feedback control of two-dimensional aeroelastic system. *Journal of Guidance, Control and Dynamics*, 29(3):762–767, 2006.

[Bar06b]   P. Baranyi. Tensor-product model-based control of two-dimensional aeroelastic system. *Journal of Guidance, Control, and Dynamics*, 29(2):391–400, 2006.

[BBC94]   R.T. Bupp, D.S. Bernstein, and V.T. Coppola. Vibration suppression of multi-modal translational motion using a rotational actuator. In *Proceedings of the 33rd IEEE Conference on Decision and Control, 1994*, volume 4, pages 4030–4034, Orlando, Florida, USA, 1994.

[BBC95]   R.T. Bupp, D.S. Bernstein, and V.T. Coppola. A benchmark problem for nonlinear control design: Problem statement, experimental testbed, and passive nonlinear compensation. In *Proceedings of the 1995 American Control Conference*, volume 6, pages 4363–4367, Seattle, Washington, USA, 1995.

[BBC98]   R.T. Bupp, D.S. Bernstein, and V.T. Coppola. A benchmark problem for nonlinear control design. *International Journal of Robust and Nonlinear Control*, 8(4-5):307–310, 1998.

[BBK89]   S. Boyd, V. Balakrishnan, and P. Kabamba. A bisection method for computing the H_∞ norm of a transfer matrix and related problems. *Mathematics of Control, Signals, and Systems (MCSS)*, 2(3):207–219, 1989.

[BEGFB94]   S. Boyd, L. El Ghaoui, E. Feron, and V. Balakrishnan. *Linear matrix inequalities in system and control theory*, volume 15. Society for Industrial Mathematics, 1994.

[Ber98]   D.S. Bernstein. Special issue: A nonlinear benchmark problem. *International Journal of Robust and Nonlinear Control*, 8:305–457, 1998.

[BG96]   J.J. Block and H. Gilliatt. Active Control of an Aeroelastic Structure. Master's thesis, Texas A&M University, 1996.

[BKPH04]   P. Baranyi, P. Korondi, R.J. Patton, and H. Hashimoto. Trade-off between approximation accuracy and complexity for TS fuzzy models. *Asian Journal of Control*, 6(1):21–33, 2004.

[BKT09]   P. Baranyi, P. Korondi, and K. Tanaka. Parallel distributed compensation based stabilization of a 3-DOF RC helicopter: A tensor product transformation based approach. *Journal of Advanced Computational Intelligence and Intelligent Informatics*, 13:25–34, 2009.

[BL91]   E.K. Blum and L.K. Li. Approximation theory and feedforward networks. *Neural Networks*, 4(4):511–515, 1991.

[BPK+07]   P. Baranyi, Z. Petres, P. Korondi, Y. Yam, and H. Hashimoto. Complexity relaxation of the tensor product model transformation for higher dimensional problems. *Asian Journal of Control*, 9(2):195–200, 2007.

[BS98]   J.J. Block and T.W. Strganac. Applied active control for a nonlinear aeroelastic structure. *Journal of Guidance, Control, and Dynamics*, 21(6):838–845, 1998.

[BSU93]   R. Bambang, E. Shimemura, and K. Uchida. Mixed H_2/H_∞ control with pole placement: State feedback case. In *Proceedings of the 1993 American Control Conference*, pages 2777–2779, San Francisco, California, USA, 1993.

[BSV06]   P. Baranyi, L. Szeidl, and P. Várlaki. Numerical reconstruction of the HOSVD based canonical form of polytopic dynamic models. In *Proceedings of the 10th International Conference on Intelligent Engineering Systems*, pages 196–201, London, UK, June 26–28, 2006.

[BSVY06]   P. Baranyi, L. Szeidl, P. Várlaki, and Y. Yam. Definition of the HOSVD based canonical form of polytopic dynamic models. In *Proceedings of the 2006 IEEE International Conference on Mechatronics*, pages 660–665, Budapest, Hungary, July 3–5, 2006.

[BTYP03]   P. Baranyi, D. Tikk, Y. Yam, and R.J. Patton. From differential equations to PDC controller design via numerical transformation. *Computers in Industry*, 51(3):281–297, 2003.

[BVK05]   P. Baranyi and A.R. Várkonyi-Kóczy. TP transformation based dynamic system modeling for nonlinear control. *IEEE Transactions on Instrumentation and Measurement*, 54(6):2191–2203, 2005.

[BY89]   S. Boyd and Q. Yang. Structured and simultaneous Lyapunov functions for system stability problems. *International Journal of Control*, 49(6):2215–2240, 1989.

[BY00]   P. Baranyi and Y. Yam. Fuzzy rule base reduction. In D. Ruan and E.E. Kerre, editors, *Fuzzy IF-THEN Rules in Computational Intelligence: Theory and Applications*, chapter 7, pages 135–160. Kluwer, 2000.

[BY06]   P. Baranyi and Y. Yam. Case study of the TP-model transformation in the control of a complex dynamic model with structural nonlinearity. *IEEE Transactions on Industrial Electronics*, 53(3):895–904, 2006.

[BYVK$^+$02] P. Baranyi, Y. Yam, A.R. Várkonyi-Kóczy, R.J. Patton, P. Michelberger, and M. Sugiyama. SVD-based complexity reduction to TS fuzzy models. *IEEE Transactions on Industrial Electronics*, 49(2):433–443, 2002.

[BYVKP03] P. Baranyi, Y. Yam, A.R. Várkonyi-Kóczy, and R.J. Patton. SVD-based reduction to MISO TS models. *IEEE Transactions on Industrial Electronics*, 50(1):232–242, 2003.

[Cas95] J.L. Castro. Fuzzy logic controllers are universal approximators. *IEEE Transactions on Systems, Man and Cybernetics*, 25(4):629–635, 1995.

[CG96] M. Chilali and P. Gahinet. H$_\infty$ design with pole placement constraints: An LMI approach. *IEEE Transactions on Automatic Control*, 41(3):358–367, 1996.

[Chu10] S. Chumalee. *Robust gain-scheduled H$_\infty$ control for unmanned aerial vehicles*. PhD thesis, Cranfield University, Cranfield, United Kingdom, 2010.

[CW09] S. Chumalee and J. Whidborne. LPV autopilot design of a Jindivik UAV. AIAA Paper 2009-5753, August 2009.

[Cyb89] G. Cybenko. Approximation by superpositions of a sigmoidal function. *Mathematics of Control, Signals, and Systems (MCSS)*, 2(4):303–314, 1989.

[DCSS78] E.H. Dowell, H.C. Curtis, R.H. Scanlan, and F. Sisto. *A Modern Course in Aeroelasticity*. Sijthoff & Noordhoff (Alphen aan den Rijn), 1978.

[Dep88] E.F. Deprettere. *SVD and Signal Processing: Algorithms, Applications and Architectures*. North-Holland Publishing Co., Amsterdam, The Netherlands, 1988.

[DGKF89] J.C. Doyle, K. Glover, P.P. Khargonekar, and B.A. Francis. State-space solutions to standard H$_2$ and H$_\infty$ control problems. *IEEE Transactions on Automatic Control*, 34(8):831–847, 1989.

[DL97] L. De Lathauwer. *Signal Processing Based on Multilinear Algebra*. PhD thesis, Katholieke Universiteit Leuven, Belgium, 1997.

[DLDMV94] L. De Lathauwer, B. De Moor, and J. Vandewalle. Blind source separation by higher order singular value decomposition. In *Proceedings of the European Signal Processing Conference, 1994*, volume 1, pages 175–178, Edinburgh, Scotland, UK, 1994.

[DLDMV97] L. De Lathauwer, B. De Moor, and J. Vandewalle. Dimensionality reduction in higher-order-only ICA. In *Proceedings of the IEEE Signal Processing Workshop on Higher-Order Statistics, 1997*, pages 316–320, Banff, Alberta, Canada, 1997.

[DLDMV00] L. De Lathauwer, B. De Moor, and J. Vandewalle. A multilinear singular value decomposition. *SIAM Journal on Matrix Analysis and Applications*, 21(4):1253–1278, 2000.

[EB02] A. Edelmayer and J. Bokor. Optimal H$_2$/H$_\infty$ scaling for sensitivity optimization of detection filters. *International Journal of Robust and Nonlinear Control*, 12(8):749–760, 2002.

[Ede87]   H. Edelsbrunner. *Algorithms in Combinatorial Geometry*, volume 10. Springer, 1987.

[EHR94]   A. El Hajjaji and A. Rachid. Explicit formulas for fuzzy controller. *Fuzzy Sets and Systems*, 62(2):135–141, 1994.

[EOSR99]   G. Escobar, R. Ortega, and H. Sira-Ramirez. Output-feedback global stabilization of a nonlinear benchmark system using a saturated passivity-based controller. *IEEE Transactions on Control Systems Technology*, 7(2):289–293, 1999.

[FAG95]   E. Feron, P. Apkarian, and P. Gahinet. S-procedure for the analysis of control systems with parametric uncertainties via parameter-dependent Lyapunov functions. In *Proceedings of the 1995 American Control Conference*, volume 1, pages 968–972, Seattle, Washington, USA, 1995.

[Fun55]   Y.C. Fung. *An Introduction to the Theory of Aeroelasticity*. John Wiley, Chichester, 1955.

[GA94]   P. Gahinet and P. Apkarian. A linear matrix inequality approach to H_∞ control. *International Journal of Robust and Nonlinear Control*, 4(4):421–448, 1994.

[GAC94]   P. Gahinet, P. Apkarian, and M. Chilali. Affine parameter-dependent Lyapunov functions for real parametric uncertainty. In *Proceedings of the 33rd IEEE Conference on Decision and Control, 1994*, volume 3, pages 2026–2031, Orlando, Florida, USA, 1994.

[Gah94]   P. Gahinet. Explicit controller formulas for LMI-based H_∞ synthesis. In *Proceedings of the 1994 American Control Conference*, volume 3, pages 2396–2400, Baltimore, Maryland, USA, 1994.

[GB99]   P. Gáspár and J. Bokor. Progress in system and robot analysis and control design, 1999.

[GG00]   I. Grattan-Guinness. A sideways look at Hilbert's twenty-three problems of 1900. *Notices of the AMS*, 47(7):752–757, 2000.

[GK65]   G. Golub and W. Kahan. Calculating the singular values and pseudo-inverse of a matrix. *Journal of the Society for Industrial and Applied Mathematics: Series B, Numerical Analysis*, pages 205–224, 1965.

[GL94]   P. Gahinet and A.J. Laub. Reliable computation of γ_{opt} in singular H_∞ control. In *Proceedings of the 33rd IEEE Conference on Decision and Control, 1994*, volume 2, pages 1527–1532, Orlando, Florida, USA, 1994.

[GNLC95]   P. Gahinet, A. Nemirovski, A. Laub, and M. Chilali. LMI control toolbox user's guide. *The Math. Works Inc.*, 1995.

[Gra00]   J. Gray. The Hilbert problems 1900–2000. *Newsletter*, 36(10-13):1–1, 2000.

[GSB02]   P. Gáspár, I. Szászi, and J. Bokor. Design of robust controllers for active vehicle suspensions. In *Proceedings of the 15th IFAC World Congress*, pages 1473–1478, Barcelona, Spain, 2002.

[GSB04]   P. Gáspár, I. Szászi, and J. Bokor. The design of a combined control
            structure to prevent the rollover of heavy vehicles. *European Journal
            of Control*, 10(2):148–162, 2004.

[GVL96]   G.H. Golub and C.F. Van Loan. *Matrix Computations*, volume 3.
            Johns Hopkins University Press, 1996.

[GWGL11]   W. Gai, H. Wang, T. Guo, and D. Li. Modeling and LPV flight con-
            trol of the canard rotor/wing unmanned aerial vehicle. In *Proceedings
            of the 2nd International Conference on Artificial Intelligence, Man-
            agement Science and Electronic Commerce (AIMSEC), 2011*, pages
            2187–2191, 2011.

[HB76]   H. Horisberger and P. Belanger. Regulators for linear time, varying
            plants with uncertain parameters. *IEEE Transactions on Automatic
            Control*, 21(5):705–708, 1976.

[Hil00]   D. Hilbert. Mathematische probleme. *2nd International Congress of
            Mathematican*, 1900. Paris, France.

[HL96]   Y. Huang and W.M. Lu. Nonlinear optimal control: Alternatives to
            Hamilton-Jacobi equation. In *Proceedings of the 35th IEEE Confer-
            ence on Decision and Control, 1996*, volume 4, pages 3942–3947,
            Kobe, Japan, 1996.

[HSW89]   K. Hornik, M. Stinchcombe, and H. White. Multilayer feedforward
            networks are universal approximators. *Neural Networks*, 2(5):359–
            366, 1989.

[IDLAVH08] M. Ishteva, L. De Lathauwer, P. Absil, and S. Van Huffel. Dimension-
            ality reduction for higher-order tensors: Algorithms and applications.
            International Journal of Pure and Applied Mathematics, 42(3):337,
            2008.

[IKM11]   Š. Ileš, F. Kolonić, and J. Matuško. Linear matrix inequalities based
            H_∞ control of gantry crane using tensor product transformation. In
            Proceedings of the 18th International Conference on Process Control,
            2011.

[IMK11]   Š. Ileš, J. Matuško, and F. Kolonić. Tensor product transformation
            based speed control of permanent magnet synchronous motor drives.
            In *Proceedings of the 17th International Conference on Electrical
            Drives and Power Electronics, EDPE 2011 (5th Joint Slovak-Croatian
            Conference)*, 2011.

[IS94]   T. Iwasaki and R.E. Skelton. All controllers for the general H_∞ con-
            trol problem: LMI existence conditions and state space formulas. *Au-
            tomatica*, 30(8):1307–1317, 1994.

[JFK96]   M. Jankovic, D. Fontaine, and P.V. Kokotovic. TORA example:
            Cascade- and passivity-based control designs. *IEEE Transactions on
            Control Systems Technology*, 4(3):292–297, 1996.

[JK00]   S.M. Joshi and A.G. Kelkar. Passivity-based robust control with ap-
            plication to benchmark active controls technology wing. *Journal of
            Guidance, Control and Dynamics*, 23(5):938–947, 2000.

[Kap77]   I. Kaplansky. *Hilbert's Problems*. University of Chicago, 1977.

[Kar84a]   N. Karmarkar. A new polynomial-time algorithm for linear programming. In *Proceedings of the 16th Annual ACM Symposium on Theory of Computing*, pages 302–311, 1984.

[Kar84b]   N. Karmarkar. A new polynomial-time algorithm for linear programming. *Combinatorica*, 4(4):373–395, 1984.

[KKR93]   I. Kaminer, P.P. Khargonekar, and M.A. Rotea. Mixed H_2/H_∞ control for discrete-time systems via convex optimization. *Automatica*, 29(1):57–70, 1993.

[KKS97a]   J. Ko, A.J. Kurdila, and T.W. Strganac. Nonlinear control of a prototypical wing section with torsional nonlinearity. *Journal of Guidance, Control and Dynamics*, 20(6):1180–1189, 1997.

[KKS97b]   J. Ko, A.J. Kurdila, and T.W. Strganac. Nonlinear control theory for a class of structural nonlinearities in a prototypical wing section. In *Proceedings of the 35th Aerospace Sciences Meeting, American Institute of Aeronautics and Astronautics, AIAA Paper No. 97-0580*, Reno, Nevada, USA, 1997.

[Kol57]   A.N. Kolmogorov. On the representation of continuous functions of many variables by superposition of continuous functions of one variable and addition. In *Doklady Akademii Nauk SSSR*, volume 114, pages 953–956, 1957.

[Kos92]   B. Kosko. Fuzzy systems as universal approximators. In *Proceedings of the 1st IEEE International Conference on Fuzzy Systems, 1992*, pages 1153–1162, San Diego, California, USA, 1992.

[KP06]   F. Kolonic and A. Poljugan. Experimental control design by TP model transformation. In *Proceedings of the 2006 IEEE International Conference on Mechatronics*, pages 666–671, Budapest, Hungary, 2006.

[KPP06]   F. Kolonic, A. Poljugan, and I. Petrovic. Tensor product model transformation-based controller design for gantry crane control system—An application approach. *Acta Polytechnica Hungarica*, 3(4):95–112, 2006.

[KR91]   P.P. Khargonekar and M.A. Rotea. Mixed H_2/H_∞ control: A convex optimization approach. *IEEE Transactions on Automatic Control*, 36(7):824–837, 1991.

[KSK98]   J. Ko, T.W. Strganac, and A.J. Kurdila. Stability and control of a structurally nonlinear aeroelastic system. *Journal of Guidance, Control and Dynamics*, 21:718–725, 1998.

[KSK99]   J. Ko, T.W. Strganac, and A.J. Kurdila. Adaptive feedback linearization for the control of a typical wing section with structural nonlinearity. *Nonlinear Dynamics*, 18(3):289–301, 1999.

[KSSK07]   Y. Kunii, B. Solvang, G. Sziebig, and P. Korondi. Tensor product transformation based friction model. In *Proceedings of the 11th International Conference on Intelligent Engineering Systems, 2007*, pages 259–264, 2007.

[LC06]   C.M. Lin and W.L. Chin. Adaptive decoupled fuzzy sliding-mode control of a nonlinear aeroelastic system. *Journal of Guidance, Con-

trol and Dynamics, 29(1):206–209, 2006.

[LCN⁺10]   B.H. Lee, J. Choo, S. Na, P. Marzocca, and L. Librescu. Sliding mode robust control of supersonic three degrees-of-freedom airfoils. *International Journal of Control, Automation and Systems*, 8(2):279–288, 2010.

[LKPV11]   T. Luspay, B. Kulcsar, T. Péni, and I. Varga. Freeway ramp metering: An LPV set theoretical analysis. In *Proceedings of the 2011 American Control Conference*, pages 733–738, 2011.

[LL86]   B.H. Lee and P. LeBlanc. Flutter analysis of a two-dimensional airfoil with cubic non-linear restoring force. Aeronautical Note-36 25438, National Aeronautical Establishment, National Research Council, Ottawa, Canada, 1986.

[Lor66]   G.G. Lorentz. *Approximation of Functions*. Holt, Rinehart and Winston (New York), 1966.

[LPK12]   T. Luspay, T. Péni, and B. Kulcsar. Constrained freeway traffic control via linear parameter varying paradigms. *Control of Linear Parameter Varying Systems with Applications*, page 461, 2012.

[MDM95]   M.S. Moonen and B. De Moor. *SVD and Signal Processing III: Algorithms, Architectures, and Applications*. Elsevier Science, 1995.

[MOS98]   I. Masubuchi, A. Ohara, and N. Suda. LMI-based controller synthesis: A unified formulation and solution. *International Journal of Robust and Nonlinear Control*, 8(8):669–686, 1998.

[Mos99]   B. Moser. Sugeno controllers with a bounded number of rules are nowhere dense. *Fuzzy Sets and Systems*, 104(2):269–277, 1999.

[MS74]   G. Matsaglia and G.P.H. Styan. Equalities and inequalities for ranks of matrices. *Linear and Multilinear Algebra*, 2(3):269–292, 1974.

[Muk00]   V. Mukhopadhyay. Transonic flutter suppression control law design and wind-tunnel test results. *Journal of Guidance, Navigation, and Control*, 23(5):930–937, 2000.

[NBG08]   Sz. Nagy, P. Baranyi, and P. Gáspár. Rollover prevention of a heavy vehicle via TP model based H-infinity control design approach. *Acta Technica Jauriensis, Series Intelligentia Computatorica*, 1(3):531–546, December 2008.

[NBP08]   S. Nagy, P. Baranyi, and Z. Petres. Centralized tensor product model form. In *6th International Symposium on Applied Machine Intelligence and Informatics, 2008*, pages 189–193, 2008.

[NG94]   A. Nemirovskii and P. Gahinet. The projective method for solving linear matrix inequalities. In *Proceedings of the 1994 American Control Conference*, volume 1, pages 840–844, Baltimore, Maryland, USA, 1994.

[NK92]   H.T. Nguyen and V. Kreinovich. On approximations of controls by fuzzy systems. LIFE Chair of Fuzzy Theory TR 92-93/302, Tokyo Institute of Technology, 1992.

[NN94]   Y. Nesterov and A. Nemirovskii. *Interior-point polynomial algorithms in convex programming*. SIAM Studies in Applied Mathematics, 1994.

[NPBH09]   S. Nagy, Z. Petres, P. Baranyi, and H. Hashimoto. Computational relaxed TP model transformation: Restricting the computation to subspaces of the dynamic model. *Asian Journal of Control*, 11(5):461–475, 2009.

[OGS96]   T. O'Neil, H.C. Gilliat, and T.W. Strganac. Investigations of aeroelastic response for a system with continuous structural nonlinearities. In *Proceedings of 35th AIAA Structures, Structural Dynamics, and Materials Conference AIAA paper*, pages 96–1390, Hilton Head, South Carolina, USA, 1996.

[O'R98]   J. O'Rourke. *Computational Geometry in C*. Cambridge University Press, 1998.

[OS95]   T. O'Neil and T.W. Strganac. An experimental investigation of nonlinear aeroelastic response. In *Proceedings of 36th AIAA Structures, Structural Dynamics, and Materials Conference AIAA paper 95-1404*, pages 2043–2051, New Orleans, Louisiana, USA, 1995.

[PAL95]   S.J. Price, H. Alighanbari, and B.H.K. Lee. The aeroelastic response of a two-dimensional airfoil with bilinear and cubic structural nonlinearities. *Journal of Fluids and Structures*, 9(2):175–193, 1995.

[PBH10]   Z. Petres, P. Baranyi, and H. Hashimoto. Approximation and complexity trade-off by TP model transformation in controller design: A case study of the TORA system. *Asian Journal of Control*, 12(5):575–585, 2010.

[PBKH07]   Z. Petres, P. Baranyi, P. Korondi, and H. Hashimoto. Trajectory tracking by TP model transformation: Case study of a benchmark problem. *Industrial Electronics, IEEE Transactions on*, 54(3):1654–1663, 2007.

[PCD08]   Z. Prime, B. Cazzolato, and C. Doolan. A mixed H_2/H_∞ scheduling control scheme for a two degree-of-freedom aeroelastic system under varying airspeed and gust conditions. In *Proceedings of the AIAA Guidance, Navigation and Control Conference*, pages 18–21, Honolulu, Hawaii, USA, 2008.

[PCDS10]   Z. Prime, B. Cazzolato, C. Doolan, and T. Strganac. Linear-parameter-varying control of an improved three-degree-of-freedom aeroelastic model. *Journal of Guidance, Control and Dynamics*, 33(2):615–619, 2010.

[PD93]   A. Packard and J. Doyle. The complex structured singular value. *Automatica*, 29(1):71–109, 1993.

[PDP+10]   R.E. Precup, L.T. Dioanca, E.M. Petriu, M.B. Rădac, S. Preitl, and C.A. Dragoş. Tensor product-based real-time control of the liquid levels in a three tank system. In *Proceedings of the IEEE/ASME International Conference on Advanced Intelligent Mechatronics (AIM), 2010*, pages 768–773, 2010.

[PDP+12]   R.E. Precup, C.A. Dragoş, S. Preitl, M.B. Rădac, and E.M. Petriu. Novel tensor product models for automatic transmission system control. *IEEE Systems Journal*, 6(3):488–498, 2012.

[PH11]   R.E. Precup and H. Hellendoorn. A survey on industrial applications of fuzzy control. *Computers in Industry*, 62(3):213–226, 2011.

[PLA94]   S.J. Price, B.H.K. Lee, and H. Alighanbari. Postinstability behavior of a two-dimensional airfoil with a structural nonlinearity. *Journal of Aircraft*, 31(6):1395–1401, 1994.

[PPU⁺08]   R.E. Precup, S. Preitl, B.I. Ursache, P.A. Clep, P. Baranyi, and J.K. Tar. On the combination of tensor product and fuzzy models. In *Proceedings of the IEEE International Conference on Automation, Quality and Testing, Robotics, 2008*, volume 2, pages 48–53, 2008.

[PSG99]   L. Palkovics, A. Semsey, and E. Gerum. Roll-over prevention system for commercial vehicles—Additional sensorless function of the electronic brake system. *Vehicle System Dynamics*, 32(4-5):285–297, 1999.

[QZL⁺11]   W.W. Qin, Z.Q. Zheng, G. Liu, J.J. Ma, and W.Q. Li. Robust variable gain control for hypersonic vehicles based on LPV. *Systems Engineering and Electronics*, 33(6):1327–1331, 2011.

[RKM92]   R.H. Rand, R.J. Kinesey, and D.L. Mingori. Dynamics of spinup through resonance. *International Journal of Non-linear Mechanics*, 27(3):489–502, 1992.

[RSV11]   A. Rovid, L. Szeidl, and P. Varlaki. On tensor-product model based representation of neural networks. In *Proceedings of the 15th IEEE International Conference on Intelligent Engineering Systems (INES), 2011*, pages 69–72, 2011.

[RW11]   S.L.M.D. Rangajeeva and J.F. Whidborne. Linear parameter varying control of a quadrotor. In *Proceedings of the 6th IEEE International Conference on Industrial and Information Systems (ICIIS), 2011*, pages 483–488, 2011.

[SBS99]   Z. Szabó, J. Bokor, and F. Schipp. Identification of rational approximate models in H^∞ using generalized orthonormal basis. *IEEE Transactions on Automatic Control*, 44(1):153–158, 1999.

[Sch92]   C. Scherer. H_∞-optimization without assumptions on finite or infinite zeros. *SIAM Journal on Control and Optimization*, 30(1):143–166, 1992.

[SD91]   G. Stein and J. Doyle. Beyond singular values and loop shapes. *Journal of Guidance, Control, and Dynamics*, 14:5–16, 1991.

[SG05]   S.N. Singh and S. Gujjula. Variable structure control of unsteady aeroelastic system with partial state information. *Journal of Guidance, Control and Dynamics*, 28(3):568–573, 2005.

[SGB10]   Z. Szabó, P. Gáspár, and J. Bokor. A novel control-oriented multi-affine qLPV modeling framework. In *Proceedings of the 18th Mediterranean Conference on Control & Automation (MED), 2010*, pages 1019–1024, 2010.

[SGNB08]   Z. Szabó, P. Gáspár, S. Nagy, and P. Baranyi. TP model transformation for control-oriented qLPV modeling. *Australian Journal of Intelligent*

Information Processing Systems, 10(2):36–53, 2008.

[SHQW12]   C. Sun, Y. Huang, C. Qian, and L. Wang. On modeling and control of a flexible air-breathing hypersonic vehicle based on LPV method. *Frontiers of Electrical and Electronic Engineering*, 7(1):1–13, 2012.

[SP00]   R.C. Scott and L.E. Pado. Active control of wind-tunnel model aeroelastic response using neural networks. *Journal of Guidance, Control and Dynamics*, 23(6):1100–1108, 2000.

[Spr65]   D.A. Sprecher. On the structure of continuous functions of several variables. *Transactions of the American Mathematical Society*, 115:340–355, 1965.

[Ste92]   G.W. Stewart. On the early history of singular value decomposition. Technical Report TR–92–31, Institute for Advanced Computer Studies, University of Maryland, March 1992.

[Sug85]   M. Sugeno. An introductory survey of fuzzy control. *Information Sciences*, 36(1-2):59–83, 1985.

[SV09]   L. Szeidl and P. Várlaki. HOSVD based canonical form for polytopic models of dynamic systems. *Journal of Advanced Computational Intelligence and Intelligent Informatics*, 13(1):52–60, 2009.

[SW00]   C. Scherer and S. Weiland. Linear matrix inequalities in control. *Lecture Notes, Dutch Institute for Systems and Control, Delft, The Netherlands*, 2000. http://www.cs.ele.tue.nl/sweiland/lmi.htm.

[Tad01]   G. Tadmor. Dissipative design, lossless dynamics, and the nonlinear TORA benchmark example. *IEEE Transactions on Control Systems Technology*, 9(2):391–398, 2001.

[Tak11]   B. Takarics. *TP Model Transformation Based Sliding Mode Control and Friction Compensation*. PhD thesis, Budapest University of Technology and Economics, Budapest, Hungary, 2011.

[TBP02]   D. Tikk, P. Baranyi, and R.J. Patton. Polytopic and TS models are nowhere dense in the approximation model space. In *Proceedings of the 2002 IEEE International Conference on Systems, Man and Cybernetics*, volume 7, Yasmine Hammamet, Tunisia, 2002.

[TBP07]   D. Tikk, P. Baranyi, and R.J. Patton. Approximation properties of TP model forms and its consequences to TPDC design framework. *Asian Journal of Control*, 9(3):221–231, 2007.

[TBnt]   B. Takarics and P. Baranyi. Friction compensation in TP model form—Aeroelastic wing as an example system. *Acta Polytechnica Hungarica*, in print.

[TIW96]   K. Tanaka, T. Ikeda, and H.O. Wang. Design of fuzzy control systems based on relaxed LMI stability conditions. In *Proceedings of the 35th IEEE Conference on Decision and Control, 1996*, volume 1, pages 598–603, Kobe, Japan, 1996.

[TOP11]   E.S. Tognetti, R.C.L.F. Oliveira, and P.L.D. Peres. An LMI-based approach to static output feedback stabilization of TS fuzzy systems. In *Proceedings of the 18th IFAC World Congress*, pages 12593–12598, Milano, Italy, 2011.

[TOW04]   K. Tanaka, H. Ohtake, and H.O. Wang. A practical design approach to stabilization of a 3-DOF RC helicopter. *IEEE Transactions on Control Systems Technology*, 12(2):315–325, 2004.

[TS90]   K. Tanaka and M. Sugeno. Stability analysis of fuzzy systems using Lyapunov's direct method. In *Proceedings of the North American Fuzzy Information Processing Society*, pages 133–136, 1990.

[TS92]   K. Tanaka and M. Sugeno. Stability analysis and design of fuzzy control systems. *Fuzzy Sets and Systems*, 45(2):135–156, 1992.

[TTW98]   K. Tanaka, T. Taniguchi, and H.O. Wang. Model-based fuzzy control of TORA system: Fuzzy regulator and fuzzy observer design via LMIs that represent decay rate, disturbance rejection, robustness, optimality. In *Proceedings of the 7th IEEE International Conference on Fuzzy Systems, 1998*, volume 1, pages 313–318, Alaska, USA, 1998.

[TW01]   K. Tanaka and H.O. Wang. *Fuzzy Control Systems Design and Analysis: A Linear Matrix Inequality Approach*. Wiley-Interscience, 2001.

[Vac91]   R.J. Vaccaro. *SVD and Signal Processing II: Algorithms, Analysis and Applications*. Elsevier Science, 1991.

[Wan92]   L.X. Wang. Fuzzy systems are universal approximators. In *Proceedings of the 1st IEEE International Conference on Fuzzy Systems, 1992*, pages 1163–1170, San Diego, California, USA, 1992.

[Was97]   M.R. Waszak. Robust multivariable flutter suppression for the benchmark active control technology (BACT) wind-tunnel model. *Journal of Guidance, Control, and Dynamics*, 24(1):147–143, 1997.

[WB92]   B. Wie and D. Bernstein. Benchmark problems for robust control design. *Journal of Guidance, Control, and Dynamics*, 15(5):1057–1059, 1992.

[WBC96]   C.J. Wan, D.S. Bernstein, and V.T. Coppola. Global stabilization of the oscillating eccentric rotor. *Nonlinear Dynamics*, 10(1):49–62, 1996.

[WTG95]   H.O. Wang, K. Tanaka, and M. Griffin. Parallel distributed compensation of nonlinear systems by Takagi-Sugeno fuzzy model. In *Proceedings of the International Joint Conference of the 4th IEEE International Conference on Fuzzy Systems and the 2nd International Fuzzy Engineering Symposium, 1995*, volume 2, pages 531–538, Yokohama, Japan, 1995.

[XS00]   W.H. Xing and S.N. Singh. Adaptive output feedback control of a nonlinear aeroelastic structure. *Journal of Guidance, Control and Dynamics*, 23(6):1109–1116, 2000.

[Yam97]   Y. Yam. Fuzzy approximation via grid point sampling and singular value decomposition. *IEEE Transactions on Systems, Man, and Cybernetics, Part B: Cybernetics*, 27(6):933–951, 1997.

[YBY99]   Y. Yam, P. Baranyi, and C.T. Yang. Reduction of fuzzy rule base via singular value decomposition. *IEEE Transactions on Fuzzy Systems*, 7(2):120–132, 1999.

[YND95]   P.M. Young, M.P. Newlin, and J.C. Doyle. Let's Get Real. *Robust*

Control Theory, 66:143–174, 1995.

[YSW02]   W. Yim, S.N. Singh, and W. Wells. Nonlinear control of a prototypical aeroelastic wing section: State-dependent Riccati equation method. In *Proceedings of the 4th International Conference on Nonlinear Problems in Aviation and Aerospace (ICNPAA'02)*, pages 543–550, Melbourne, Florida, USA, 2002.

[YYB03]   Y. Yam, C.T. Yang, and P. Baranyi. Singular value-based fuzzy reduction with relaxed normality condition. In J. Casillas, O. Cordon, F. Herrera, and L. Magdalena, editors, *Interpretability Issues in Fuzzy Modeling*, volume 128 of *Studies in Fuzziness and Soft Computing*, pages 325–354. Springer Verlag, 2003.

[ZY90]   L.C. Zhao and Z.C. Yang. Chaotic motions of an airfoil with nonlinear stiffness in incompressible flow. *Journal of Sound and Vibration*, 138(2):245–254, 1990.

Index

A

B

Printed and bound by CPI Group (UK) Ltd, Croydon, CR0 4YY

18/10/2024

01776261-0003